T0192734

Science and Fiction

Science and Fiction – A Springer Series

This collection of entertaining and thought-provoking books will appeal equally to science buffs, scientists and science-fiction fans. It was born out of the recognition that scientific discovery and the creation of plausible fictional scenarios are often two sides of the same coin. Each relies on an understanding of the way the world works, coupled with the imaginative ability to invent new or alternative explanations—and even other worlds. Authored by practicing scientists as well as writers of hard science fiction, these books explore and exploit the borderlands between accepted science and its fictional counterpart. Uncovering mutual influences, promoting fruitful interaction, narrating and analyzing fictional scenarios, together they serve as a reaction vessel for inspired new ideas in science, technology, and beyond.

Whether fiction, fact, or forever undecidable: the Springer Series "Science and Fiction" intends to go where no one has gone before!

Its largely non-technical books take several different approaches. Journey with their authors as they

- Indulge in science speculation—describing intriguing, plausible yet unproven ideas;
- Exploit science fiction for educational purposes and as a means of promoting critical thinking;
- Explore the interplay of science and science fiction—throughout the history of the genre and looking ahead;
- Delve into related topics including, but not limited to: science as a creative process, the limits of science, interplay of literature and knowledge;
- Tell fictional short stories built around well-defined scientific ideas, with a supplement summarizing the science underlying the plot.

Readers can look forward to a broad range of topics, as intriguing as they are important. Here just a few by way of illustration:

- Time travel, superluminal travel, wormholes, teleportation
- Extraterrestrial intelligence and alien civilizations
- Artificial intelligence, planetary brains, the universe as a computer, simulated worlds
- Non-anthropocentric viewpoints
- Synthetic biology, genetic engineering, developing nanotechnologies
- Eco/infrastructure/meteorite-impact disaster scenarios
- Future scenarios, transhumanism, posthumanism, intelligence explosion
- Virtual worlds, cyberspace dramas
- Consciousness and mind manipulation

More information about this series at http://www.springer.com/series/11657

Thomas Eversberg

The Moon Hoax?

Conspiracy Theories on Trial

 Springer

Thomas Eversberg
German Space Agency
Bonn, Germany

Translated into English by Markus Josef Geiss and Jordan Barr Holquist

Translation from the German language edition: "Hollywood im Weltall" by Thomas Eversberg. © Springer-Verlag GmbH 2013.

ISSN 2197-1188   ISSN 2197-1196 (electronic)
Science and Fiction
ISBN 978-3-030-05459-5   ISBN 978-3-030-05460-1 (eBook)
https://doi.org/10.1007/978-3-030-05460-1

Library of Congress Control Number: 2018966595

This Springer imprint is published by the registered company Springer Nature Switzerland AG
The registered company address is: Gewerbestrasse 11, 6330 Cham, Switzerland

For Bine

Foreword

July 20, 1969: The first humans land on the Moon. The older generations among us witnessed this event, whereas the younger generations have learned about it from history books. A historic event! A giant leap for mankind! Or maybe not? Could the event that was broadcast to millions of TVs possibly have been staged: is this a case of a "Moon Hoax"? There have been persistent rumors that the US space agency, NASA, tricked everyone with smoke and mirrors, and that all of the technological advances were completely made up. In the 1970s, the lies about the faked Moon landings were born. And nowadays, in the time of the Internet, where everyone is not only a consumer of media but can also easily be the author and distributor of information, these lies are being spread constantly. Other conspiracy theories, too, are once more rearing their ugly heads, all on the World Wide Web.

Our human insistence on doubting and questioning events, claims, and alleged facts is actually a positive aspect of our culture. These are inevitable requirements to be able to understand correlations, to classify things, and to increase knowledge. But where is the border between common sense and scientific thinking on the one hand, and lack of understanding, confusion, and ideological delusion on the other? What are we willing to accept as true and what remains incompatible with our worldview?

Thomas Eversberg, PhD in astrophysics and an active professional in space management, deals with the arguments of those who would question the trips to the Moon in this book, *The Moon Hoax?—Conspiracy Theories on Trial.* By taking these arguments seriously and then confronting them with solid logic, his analysis turns into a unique lesson. With penetrating clarity, he uses a tool from the philosophy of science called Occam's Razor: According to this theory, the hypothesis describing a phenomenon with the fewest assumptions should be preferred. Those requiring an unnecessarily large number of assumptions can be discarded as too complex (in a figure of speech: sliced off by the razor blade).

It is the consistent application of this principle, a common practice in scientific work and rational thinking, that makes Eversberg's analysis significant far beyond the topic at hand. The author shows not only the flaws in the arguments of the Moon landing opponents—he shows more generally how serious arguments can be distinguished from fantasies. Anyone who reads this book will become much more capable of navigating the vast flood of information to be found in modern media without running the risk of being defrauded.

Spektrum der Wissenschaft Verlagsgesellschaft mbH   Uwe Reichert
Heidelberg, Germany

Acknowledgement

Who would have thought that a little boy enthusiastic about space would turn into an astrophysicist that deals with the Moon landings 40 years later? I am indebted to my grandmother, Ruth Wendland, who sensitively promoted my interests, as did my father, Karl-Werner Eversberg. I would like to give a heartfelt thanks to him and my mother, Karin Eversberg, who gave me total freedom and who always supported my curiosity and enthusiasm. Many friends inspired me to give talks about the Moon landings and therefore contributed significantly to this book through their various questions and thoughts. This particularly applies to a few people that I would like to mention here. The many discussions I had with Andreas Boeckh instigated by his interest in science and the Moon, whether at home or in the Swedish mountains, gave rise to many of my approaches in this book. If he was ever bored by these weird thoughts of mine, he politely never let me notice. Norbert Reinecke deserves special thanks and respect for his critical questions and comments about my endeavors as an astronomer, and his support during rough times. I would like to thank Klaus Vollmann for our joint scientific discussions, his dedication to scientific accuracy, and the work at our observatory for many years; even more so, because this work is often very tiring and sometimes not the most enjoyable. Moreover, I would like to thank Anke Gödersmann and Dieter Schaade for their inspiring discussions while sharing delicious meals with me for many years. I further want to thank my uncle, Abdelali Aouati, for the constant motivation and his unprecedented optimism. And I thank my good friend Britta Schlörscheidt for motivating me during the writing of the book. Also, thank you to Martina Mechler from Springer Spektrum for her great help in creating this book. This also applies to my editor, Vera Spillner, who made a major contribution by asking professional and critical questions while giving me careful recommendations. I also want to thank my editor, Angela Lahee, for making the English edition possible. Last but not least, I would like to thank my wife Sabine for her affection and endless patience. I especially appreciate her for these when I descend into logical questions, float off on stellar winds, or when I'm on the Moon once again.

Fig. 1 The *Saturn V* Moon rocket with *Apollo 11* on its way to the Moon. This rocket, the most powerful machine ever built, had a weight of almost 3000 metric tons, a total height of 111 meters, and developed a thrust of 3500 tons, or 160 million horse power. Photo: NASA. No.: AP11-KSC-69PC-442

Contents

1

Prologue: The Conspiracy of the Faked Moon Landings

A couple of years ago, many friends of mine asked whether I actually believed in the American Moon landings of the 60s and 70s. I was not the least bit surprised about this question. When I was a child, I unsurprisingly paid close attention to the Moon landings and earnestly painted rockets in school. We kids knew the names of our heroes by heart and fought over the question of who would be the best astronaut. For some reason I thought Jim Lovell from Apollo 8 was a particularly good astronaut, but Frank Borman also wasn't bad. My most envied possession was my Apollo-Quartet collectible card game, and to feed my enthusiasm, my grandmother sent me a picture book about the path to the Moon landings that I absolutely adored. The Moon landings are a part of my childhood and they were the grounds for my passion about space, as well as for my technical and scientific interests. Lastly, these missions were the foundations as to why I would become an astronomer and why I currently work in aerospace engineering management (Fig. 1.1). And now here comes this question!

I had, of course, realized that for some time in the media, and especially on the internet, there was a massive amount of doubt about the authenticity of the Moon landings. Based on irritating photographs, people claimed that humans had never actually been to the Moon, and that all of the reports, films, and results were one enormous trick performed on the entire world. I was only partially aware of these conspiracies and never really paid any attention to them.

But then my friends, intelligent people who are able to distinguish between serious arguments and fantasies, came to me with these questions. They were unsettled by the various claims from the so-called "Moon landing deniers"

© Springer Nature Switzerland AG 2019
T. Eversberg, *The Moon Hoax?*, Science and Fiction,
https://doi.org/10.1007/978-3-030-05460-1_1

Fig. 1.1   Earthrise. Photo: NASA/E. Cernan. No.: AS17-152-23274

and wondered whether there was anything true with their theories. For example, in images of the Moon published by NASA, which are freely available on the internet, something seems wrong with the shadows.[1] They don't appear to run parallel to one another, even though the Sun, which should be the only source of light, is so far away! Caught off guard, I investigated some other arguments. And indeed, something was wrong in other images as well. They appeared confusing and seemingly contradictory, and thus, my interest was aroused.

Generally, I find critical thinkers to be very pleasant, especially those who question any statement and don't just blindly accept such assertions as the

[1] Comprehensive sources for image and film documents are the NASA History Office (http://history.nasa.gov) and the Apollo Archive of Kipp Teague (http://www.apolloarchive.com).

truth. Because of that, and because I like to get to the bottom of things that are unclear to me, I was no longer able to ignore the claims from conspiracy theorists. Without addressing these assertions, I would not have been true to myself or to my skeptical friends, even more so since I am an analytical person who has enjoyed an extensive education in the natural sciences. In order to view the Moon landings from the proper perspective, we need to take into account their enormity and audacity. Humans actively worked towards leaving their home planet, an unprecedented endeavor symbolizing a break in human history; an event of the century. This was even more true because of the considerable risks taken by the astronauts during this project. Humans had first discovered how to use flying machines just 50 years earlier, and rocket technology was not even 20 years old when the Americans decided to fly to the Moon. And because some of the necessary technology for their mission didn't even exist, the idea of making this spectacular leap a reality in only ten years was simply unimaginable. Yes, it was absurd! Was everything just a lie after all?

In the year 2009, the news caught my attention that NASA had not been able to find the original photographs of the first landing on the Moon for three years. At first, I spontaneously classified this as fear mongering by some uninformed circles, but then I was more than a little irritated when the story turned out to be true. Even after intense search campaigns they were not able to recover the 45 magnetic tapes. Every rational human being has to pose the question: How in the world can something like this even happen?

Therefore, it is quite understandable if critical thinkers don't trust this entire story. Regardless of whether one is in politics or business—lies have been and will remain to be a part of every society. We are lied to in order to go to war or in order to get more money out of our pockets. The public outcry after any of these lies is great, but then we are lied to again only a few years later. Memories are fleeting after all! Even psychologists agree with the fact that lies are an essential part of the human condition and only with them are humans able to manage their everyday lives. To this extent, it makes complete sense that skepticism and detailed analyses are so close to a scarcely imaginable event. As such, critical attention to detail and a verification of facts that are presented is highly desirable—these are, of course, my daily bread and butter as a physicist. These habits are generally good to practice in your daily life, even though you can't be an expert in every field. With that and with my 40 years of work experience in space flight, I have become the first person to talk to for all of my friends' concerns regarding the Moon landing hoax. Step by step, I discovered unexpected pitfalls and the true complexity of these questions. The conversations with my friends resulted in an examination of the pros and cons of the Moon landings, whereas my approach in this book was purely analytical and based on logic. Now one might think that logic is a science in and of itself, and that I am not an expert in that. However, I would like to emphasize that logic is not borrowed, nor was it invented by science, but rather that every human being thinks and acts more or less logically every day. That is the only reasonable way to construct your life. The common phrase "That makes sense!" encapsulates this fairly well. Moreover, everyday relationships between things and actions are connected through reason, and therefore logic (the scientist calls this "causally" connected). We even learn this as children. For example, if I hold my fingers into fire, I will burn myself—that's logical.

With this in mind, I started to more closely investigate the individual items that serve as evidence to the assertion that humanity was fooled and that the whole story was made up from beginning to end. At the time, I had no idea that this would become such a large amount of work. It turns out that there is a large discrepancy between making an ad-hoc claim and the effort to either support or disprove this claim in a meaningful, but simple manner. Furthermore, it is not sufficient to only investigate individual critical points, but you also need to look at the nature of evidence and its historical context. This is especially important because we are dealing with a singular and significant event in history. It is also interesting to know who was first to question the authenticity of the Moon landings and whether or not we will ever go back (or go for the first time ever) to the Moon in the future. Out of my personal interest in the future of space flight, I noted down some thoughts regarding this topic—and the result is this book.

To access all the original texts and film materials I have reviewed for this book, you can use the QR-codes and URLs within this book to find them online.

2

Russians, Rockets, and Election Campaigns

The first time I was allowed to watch TV in the middle of the night was in 1969, when I was an eight-year-old boy, to watch the very first Moon landing happen live. At the time, I was completely unaware of the significance of the special event that I was watching unfold. Today, you can hardly believe what was going on in the media back then. Reports of new rocket launches and space missions were still completely new phenomena and these events were followed closely by the general public. They were all broadcast on live TV and fascinated everyone else just as much as they did me. The "Conquest of Space" had already been going on for ten years, and everyone was fairly certain that the Olympics would be held on the Moon in the year 2000. We argued amongst ourselves about how new records would have to be evaluated in reduced gravity (imagine a 500-meter javelin toss), and becoming an astronaut was THE DREAM of all young boys.[1] Anyway, it was absolutely clear that new worlds were opening up—the film "2001—A Space Odyssey," reflects this attitude very well. For most of the adult population in Germany, not just in our city, the event was so momentous that they woke up at 3 o'clock on a Monday morning (!) to see what was happening on our Moon. Most of the windows in our neighborhood were lit up. This excitement enraptured a considerable number of people all over the world, but those in North America were particularly happy with NASA. They had planned the landing time so that it occurred between noon and late afternoon on the 20th of July,

---

[1] Until then, the only woman to fly in space was the Russian, Valentina Tereschkova, and that was only a public relations stunt.

© Springer Nature Switzerland AG 2019
T. Eversberg, *The Moon Hoax?*, Science and Fiction,
https://doi.org/10.1007/978-3-030-05460-1_2

depending on where you lived in the United States. The first steps on the Moon occurred during prime-time TV (the best time for advertising: *The Moon Landing—Brought to you by Kellogg's!*) between 6:00 and 9:00 pm. For all the other inhabitants of the planet this meant a greater or lesser degree of inconvenience to their daily schedule depending on their longitude. The total viewership amounted to around 500 million people, at a time when there were far fewer TVs in the world than there are today. Sadly, most of the citizens of Earth were so poor that they didn't have the time nor the luxury to pursue something that didn't even improve their lives.[2]

When Neil Armstrong made his footprint in the lunar soil, it was the pinnacle of an extraordinary competition whose origins can be found in World War II. Germany had triggered an enormous technological revolution thanks to its war efforts. The generously funded German engineers developed new propulsion systems for weapons that could be used to destroy the enemies, and above all, their civilian populations. After the pulsed-jet engines that were used to propel the V1 unmanned flying bombs (propagandistically named "Vergeltungswaffe 1 (V1)" or "Retaliatory Weapon 1"), rocket engines for airplanes and missiles were developed and "successfully" used for the first time in history. The engineers designated these engines "Aggregate" (A1–A4), or "Assemblies" and the missiles "Vergeltungswaffe 2 (V2)," or "Retaliatory Weapon 2." With them, it was the first time that a warhead was dropped through the stratosphere at supersonic speeds on another country. This "technological milestone" was so important to the team around project manager Wernher von Braun, that men were enslaved and even killed to ensure the success of those projects.[3] After the Allied Forces freed Germany from the murderous Nazi regime, these technologies were claimed by the victors of the war. Diplomatic relations between the East and the West chilled following conclusion of World War II to the point that we now refer to this era of East-West opposition as the "Cold War."

A second decisive factor was the development of the atomic bomb and its subsequent deployment on Japan. Completely surprised by its power (the physicists around Robert Oppenheimer underestimated the explosive force of the first tests in the desert near Los Alamos National Laboratory by a factor of about 50) the military proceeded to develop rockets so that these tools of the

---

[2] Even today, the rich people of the world will claim that nothing was the same after this event; however, a farmer in Bangladesh might not share this view.

[3] The poor treatment of whole groups of people by the military would be repeated later in various atomic bomb tests of the nuclear powers. The attraction of new technologies is obviously a threat to one's morals. Success corrupts! This remains relevant for today's technical developments.

Devil could be safely launched at the enemy without the possibility of being defended against. On both sides (in the USA and in the Soviet Union) the militaries continued to develop the German V2 rocket, building upon the experience of the German engineers. Suddenly, all of the unspeakable acts that occurred throughout the development and production of the V2 rockets were forgotten (it was decided that the participants were only politically naïve Nazi-supporters) and those who worked on them were welcomed with open arms into the East and the West alike.[4] When the Soviets triggered the "Sputnik-Shock" with the successful launch of their first satellite, it overtly suggested that the East had strategic superiority to the militaries of the West, contrary to what people previously thought. Outlandish scenarios like "rockets could be launched at any moment to anywhere from space," were used as threats to secure more funding for rocket development programs. This was an exceedingly successful method for both sides.

The tense political and strategic mood at the time may be difficult to comprehend for younger generations born in a more or less peaceful Europe, but this conflict is an important part of scientific history and aids in the understanding of space travel. The fact that two countries spent tens of billions of dollars on what appeared to be a boondoggle from both a scientific and strategic point of view can only be understood when one takes the political environment and the general paranoia of the time into account.[5]

In order to get the Moon program off the ground, however, the military strategy required one more component: Competition. The two opposing social systems in the East (communism) and West (capitalism) became an essential vehicle of politics, and the opportunity of pitting the two systems against each other explains many political actions of the time (although explaining these political actions does not necessarily make them more sensible). This still includes technological competence and with it, the strategic strength of a government. The current day example of this is China. The ability to launch satellites into space became an important aspect of the competition between the two social systems. Especially true for the American politicians, space travel became an important topic for their elections. The successes in space of the Soviet Union were not only bad press for the USA, but the first satellite in Space, *Sputnik 1*, and particularly the first space flight of a man by Yuri Gagarin, caused a reaction bordering on hysteria from the

---

[4] According to SS documents, about 12,000 people were killed at the V-weapons plant in the Mittelbau-Dora plant.

[5] Senator McCarthy's anticommunist outbursts in the 1950s still resonates in today's outbursts of religious fundamentalism.

American public. That was something politicians could capitalize on, and certainly wanted to. Although in hindsight it is unwittingly suggested otherwise, Kennedy had a popularity problem during his tenure: the invasion of Cuba at the Bay of Pigs failed miserably just a month earlier. Following the "Sputnik-Shock" and the first flight of a human in space with Yuri Gagarin (Figs. 2.1 and 2.2), and while American engineers were dealing with technical problems and failures, President Kennedy was convinced by his consultants and proponents of space travel to use space exploration as a public relations vehicle (Fig. 2.3).

He was so clever that he took this advice and, in his famous speech before the Senate, he inspired and convinced the taxpayers that they would land an American on the Moon before the end of the decade. It was a first-class, public relations stroke of genius.[6] On both the American side and the Soviet side, efforts were made in this direction not only to demonstrate their respective technological superiority, but also their moral superiority. Thus began the so-called "Space Race," which was analogous to a sporting event. This was fantastic propaganda for both sides and nearly led the Soviets to a Moon landing with their new *N1* Moon rocket. The contest was fueled by the first orbital flight of Yuri Gagarin in April of 1961, and then the second orbital flight of a human by German Titov in August of the same year. It was only in February of 1962 that the Americans were finally able to complete their first manned, orbital flight by sending the astronaut John Glenn around the world. The first Americans in space, Alan Shephard and Virgil "Gus" Grissom, flew on ballistic parabolic trajectories (i.e. not entirely around the Earth) in May and July of 1961, respectively.

QR Code: President Kennedy's speech to the US Congress—tinyurl.com/y9oojq3z

With the deadline for the first flight to the Moon only nine years away, the entire space program plan that the Americans had needed to be completely revamped. Previously, NASA officials had assumed that the mission concepts

---

[6] A copy of the corresponding manuscript section can be found at http://history.nasa.gov/Apollomon/apollo5.pdf. The entire speech can be found at http://www.jfklibrary.org/Research/Ready-Reference/JFK-Speeches/Special-Message-to-the-Congress-on-Urgent-National Needs-May-25-1961.aspx. There corresponding audio file can also be found there.

Fig. 2.1   News of the first manned spaceflight by Yuri Gagarin. Photo: NASA

Fig. 2.2 Control panel of the Vostok-1 Spacecraft, which took Yuri Gagarin to orbit. Photo: NASA

could be developed elegantly and in a calm manner. Nobody even talked about the Moon. As early as 1955, construction began on the *X15* rocket plane that would be used to learn about glider technology for later space-flights. The three *X15* planes that were built could fly at speeds six times the speed of sound and up to an altitude of 62 miles. By 1968, nearly 200 flights were completed in these planes.[7] The Air Force also worked on a glider concept with the intent of developing it into a long-range bomber. This project, *Dyna-Soar*, began in 1957. But then Kennedy announced his plan to send people to the Moon as quickly as possible. Within that amount of time, an elegant and cost-effective vehicle would be impossible to develop; the only choice was to take the expensive route and use disposable missiles to send people to the Moon. *X15* flights continued until the program's funding eventually ran out in 1968. The *Dyna-Soar* program fared worse: its development was abandoned in favor of funding the Moon program (Figs. 2.4, 2.5, 2.6, 2.7, and 2.8).

---

[7] Neil Armstrong completed seven flights with the *X15*.

Fig. 2.3 President Kennedy announces to the US Congress the goal of bringing a person to the Moon before the end of the decade, May 25, 2961. Photo: NASA/US Congress

I am bringing up this topic so that the Moon landings can be viewed in the proper historical context between the *X15* and the *Space Shuttle*. With his decision to go to the Moon, the American president didn't only halt the logical developmental path in favor of a rapid, brute force approach, but he also significantly impacted space flight even ten years after the Moon landings, when the costs of the American space program came under scrutiny and the design of the *Space Shuttle* had to be scaled back. I'll come back to this subject in Chapter 16.

Many people see the Moon landing of 1969 as a singular event, like an expedition to the summit of Mount Everest. They forget, however, that after Kennedy's speech, NASA built an organization that would jump-start an entire industry, employing up to 400,000 people in its most active years, with many missions completed prior to the end goal of getting to the Moon. The *Mercury* Program gave the United States its first manned spaceflight experiences. After the groundbreaking flights of the famous *Mercury Seven* astronauts came the *Gemini* Program. The *Gemini* capsule was a spacecraft piloted by two astronauts (in the year prior, the Soviets had launched a capsule called

Fig. 2.4 *Redstone 1* with the *Mercury* Capsule for the first flight of the astronaut Alan Shephard from Launch pad 5 in Cape Canaveral, FL. 1961. Photo: NASA. No.: KSC-61C-181

Voskhod 1 into orbit with three cosmonauts on board) that had two major tasks: First, it needed to prove that rendezvous maneuvers were possible. These maneuvers, also known as in-space docking, would be necessary tasks to perform in the upcoming lunar missions. Second, it was important to learn what conditions and controls were conducive to the astronauts successfully

Fig. 2.5 Edward H. White completing the first American spacewalk during *Gemini 4*. After the Russian Alexei Leonov, he was the second person to carry out this experiment. Photo: NASA/James McDivitt. No.: S65-30431

performing outboard maneuvers. Both goals were achieved within the ten manned *Gemini* missions (*Gemini 3* to *12*). Only afterwards could the *Apollo* flights be carried out, and even then, the first Moon landing was only made on the fifth *Apollo* flight.

For the purpose of practicing each complex action and sequence prior to landing on the Moon, the Moon landing was actually simulated in a "studio" prior to the real Moon landing. It is simply impossible to send people to the Moon without having them train for every possible task they may need to perform and without thoroughly testing each component and procedure. Considering the dangers that the astronauts would be facing, to go without all

Fig. 2.6 *Gemini 7* in Earth orbit, photo taken by the crew of *Gemini 6* during a joint rendezvous maneuver. 1965. Photo: NASA. No.: S65-63220

of this preparation would simply be suicide. Prior to parts being built for any mission, the entire spacecraft and the astronauts' interactions with it are entirely planned out. This begins with a project definition, followed by a feasibility study, and concludes with the final design definitions. This means that the astronauts, the components and their functions, and the entire flight are documented on paper and subject to critical tests and controls by highly qualified engineers. Everything is laid out on paper before a single bolt is made. But because humans make mistakes, even this method can't entirely prevent

Fig. 2.7 Launch of *Apollo 7* on a Saturn-1B rocket. In this orbital flight in 1968, the new *Apollo* service and command modules for the lunar missions were tested for the first time. Photo: NASA. No.: AP7-KSC-68PC-185

Fig. 2.8 Neil Armstrong and Buzz Aldrin during training for the Moon landing in Houston. 1969. Photo: NASA. No.: AP11-S69-32245

catastrophic errors. The Apollo Program is not the only human spaceflight program that has had fatal accidents.[8]

One should keep in mind that the Moon landings were part of a complex development process that involved open discussions between managers, engineers, and technicians; and that this process was not impervious to many setbacks. It remains the greatest technical accomplishment of all time. Therefore, many people were involved in the work, who on top of that, were scattered across an entire continent.

Complex, expensive, and risky space programs that human beings entrust their lives to, require prompt and intensive exchange of information. Recklessness would be deadly! As long as public relations play a part, failures will adversely affect every space program. The death of Vladimir Komarov and the entire crew of the Space Shuttle Challenger are sad examples. From the Nobel laureate Richard Feynman at the end of the *Challenger* Accident Report to the US Congress[9]: "For a successful technology, reality must take precedence over public relations, for nature cannot be fooled."

---

[8] Valentin Bondarenko (1961); Virgil (Gus) Grissom, Edward White, Vladamir Komarov, Michael Adams (1967); Georgi Dobrowolski, Victor Pazayev, Vladislav Volkov (1971); Dick Scobee, Michael Smith, Ronald McNair, Ellison Onizuke, Judith Resnik, Gregory Jarvis, Christa McAuliffe (1986); Rick Husband, William McCool, Michael Anderson, David Brown, Kalpana Chawla, Laurel Clark, Ilan Ramon (2003). In addition, several rockets exploded on the ground or at launch. These incidents officially killed 157 people. Unofficial information names up to 600 dead.

[9] Feynman's report can be found on the internet at: http://history.nasa.gov/rogersrep/v2appf.htm

3

Proof I: The Dilemma

In short, those are just some of the details in the background of the Moon landings, a historical event that I have never questioned. I have been studying aerospace history for a while, following the lunar missions and subsequent projects (for example, Skylab, the Space Shuttle, Viking, Voyager) and after the end of the Cold War, I was happy to learn more about the Soviet program. But now people ask me whether any of it even happened! And if I say that it did, they ask me to prove it.

Whenever people asked me this, I would remind them that everyone followed it happening in newspapers, on television, and in books. Very quickly, however, I discovered the issue one runs into. I was met with the counter argument that the videos on television, the pictures in the newspapers, and the reports from the astronauts, engineers, journalists, and reporters were either manipulated, or those people were simply lied to. I would point out the lunar rocks that were brought back, but these are said to be counterfeit and simply made from lava. They would basically point out that if you can't see how everything was faked, then you're just being too naïve to see the truth. After a while, I would get annoyed, standing there looking like an ignorant fool, but I couldn't really figure out why. A strangely unsatisfactory feeling remained in my stomach, combined with frustration at the mutual lack of understanding, and I would have to ask myself: What on Earth just happened?

There can't be any productive discussion when you only confront others with mistrust. To understand something, you must be willing to listen and to understand the contents of the answers you receive. Flat out refusal to do so leads to a standstill in the discussion. In reality, however, providing proof can

© Springer Nature Switzerland AG 2019
T. Eversberg, *The Moon Hoax?*, Science and Fiction,
https://doi.org/10.1007/978-3-030-05460-1_3

be difficult and treacherous. What poof is being asked for? A scientific proof? Or a historical proof? Perhaps even scientific proof for an historical event? You would do well to clear this question up with the person asking for proof in order to provide the correct response. In the end, the discussion will be tricky and complicated if you don't do this first.

I have become accustomed to reciprocating the demand for proof with a request: *Until 1989, Berlin had a very famous wall. Please prove to me that this wall existed.* Of course, the other person will refer to the photos that exist. There are also witnesses and the remnants of the wall, which remain in Berlin to this day. Further, there are records in history books. My succinct answer to them is as follows: *The photos are fake, the witnesses were manipulated by the KGB, and the alleged remnants were later built by the Secret Service; and history books don't say a thing about the veracity of these people.* On the objection that hundreds of thousands of German citizens have seen and lived through the times when the wall was standing, I can confidently say that hundreds of thousands have simply been manipulated or they are lying. By this point, I'm sure that my conversation partner has a strangely unsatisfactory feeling in his stomach and, becoming frustrated, is asking himself, "What is going on here?!" I can relate. What should you do when you're speaking with someone who negates all of your arguments instead of discussing them openly to come closer to the truth? Nothing at all. Because this isn't really a conversation—it's a propaganda event being held in private.

In truth, my retort is quite unfair. The deceit in it lies in the demand for scientific proof for a historical event. However, scientific proof comes from the repeatability of an experiment. That, and only that, says something about the quality of scientific evidence. If I say that a ball falls to Earth when I let go of it, it's just a hypothesis. This hypothesis is only proven to have evidence of being true if this experiment can be repeated as often as desired.[1] With this, I can establish a theory of free fall that must be valid for all further cases of objects falling. History, on the other hand, is not repeatable but consists of singular events that can never occur exactly the same again (excluding physical phenomena such as the return of comets that exhibit scientifically determined repeatability). Scientific proof of historical events is simply impossible because they aren't repeatable. You can only resort to an "inductive" or a

---

[1] Strictly speaking, verifications in science are impossible, and theories can only be proven as well as possible through repeated experiments. There is no 100% certainty. A single counter-proof, however, and the theory is invalid. You can only prove theories wrong. For this reason, Einstein's theory of relativity is regularly tested with experiments in order to refute it, even if hardly any other scientist doubts this theory. If in the above example, the ball ever goes up instead of falling, my hypothesis is proven wrong.

"deductive" proof. The evidence for both types of proofs is based on observations and experiences. In an inductive proof, you use observed phenomena to obtain a general insight. In a deductive proof, you use general assumptions to suggest a more specific case. However, you can't establish absolute certainty about the truth of a historical statement with these proofs, because history is not repeatable. But you can certainly test the power of individual claims made by the disparagers of the Moon landings.

Now you could also argue that proving the Moon landings is possible by using logic. This is the correct method and as such, it's what I will use it. Logical approaches require that each assumption made is accepted by all parties. An example of a proof by logic: If A is greater than B, and if B is greater than C, then A is also greater than C. If any of the assumptions are in question or are actually false, such as "if A is greater than B," or "B is greater than C," then this proof is no longer true. By ignoring logic, you could claim that the Moon is made of cheese.

Of course, a proof would be possible if the person who doesn't believe me flew to the Moon and verified my claims himself (I'll come back to that in Chapter 14). At most, this would only be feasible for a few individuals and those who remained on Earth would still probably not believe them. Therefore, no one would be able to "prove" to everyone that we have landed on the Moon in a scientific sense. All we can do is inductively check the arguments of the Moon landing opponents for consistency, logic, and insight, and then determine whether they are credible or not.

At the end of Chapter 1, I hinted that the burden of proof of the new "evidence" lies with those making new claims. In science, the requirements for proving new theories are very high. Scientists are extraordinarily conservative, and they heavily scrutinize new theories, methods, rules, or theses before accepting them as the truth. Only when new theories withstand this level of review, modifications of previous theories are considered. This was the case for Newton, Einstein, Planck, and Heisenberg. But opponents of the Moon landings consistently claim to use scientific methods for their arguments. Therefore, it is natural to discuss the topic, in contrast to NASA, in order for the reader (and for my friends) to see the theories clearly. To do this, I will present the main claims of the lunar landing opponents and put them to the test of logic. For some theories, this is very easy to do on Earth, while others will require an application of high school-level physics. In addition, I'll discuss direct evidence to the reality of the lunar missions, which in a twist of fate, can be discovered based on the observations of the conspiracy theorists.

4

Stars are Missing in the Sky

Why is the sky blue? This seemingly simple question has become so common on physics tests that it doesn't frighten students anymore. While it's easy for a layman to understand, it is not easy to answer. In fact, a detailed answer is not simple, but requires knowledge of atomic and molecular physics. The blue color of the sky is caused by the scattering of light off air particles, or the splitting of a directed light beam in other directions. This fact has been known for a long time, but interestingly, the scattering depends on the color of the light. Blue light is more strongly scattered than green or red, and so blue light is distributed over the entire daytime sky, while all other colors are mostly allowed to transmit directly through the atmosphere. The scattering of the blue light is so strong that considerable amounts of intense sunlight are spread across the sky, rather than reaching the ground. The light of the stars is outshined by the scattered sunlight and the stars are therefore obscured during the day.

But there isn't any atmosphere on the Moon. Because of this, light from outer space is not scattered and reaches the surface of the Moon without obstruction. Of course, sunlight can still blind your eyes and even be dangerous for astronauts since the ultraviolet light is not filtered out by an atmosphere. Appropriate protective measures on the astronauts' helmets, such as shades and sunglass lenses, are therefore necessary to protect their eyes. However, the stars are not hidden by a blue sky and are always visible. The sky is not blue, but always black. It does not matter whether the sun is in the sky or not—if your eyes are not blinded, stars must be visible from the Moon at all times.

With these facts of physics in mind, I'll begin with the most popular argument for the lunar landing conspiracies. In the photos of the Moon landings

© Springer Nature Switzerland AG 2019
T. Eversberg, *The Moon Hoax?*, Science and Fiction,
https://doi.org/10.1007/978-3-030-05460-1_4

Fig. 4.1 Buzz Aldrin looks at the *Apollo 11* lunar lander *Eagle*. Photo: NASA/N. Armstrong. No.: AS11-40-5948

every single star is missing from the sky. Whichever image of the lunar missions you look at, the stars are always missing. The simple argument presented is: **The Moon landing scenes were filmed in a studio and people forgot to install artificial stars on the ceiling of the studio** (Fig. 4.1).

At first glance, this assertion is as simple as it is captivating. When we look at the sky on a clear night, we see the stars. And the darker the surroundings of your location, the more stars you can see. Shouldn't the night sky visible from the Moon be fantastic, since it has no atmospheric disturbances? As clear and simple as this argument seems, it is anything but that. On the contrary, the claim that studio technicians forgot to install lamps on the ceiling has far more complex consequences than can be seen at first.[1] To help with our answer, let's take a brief excursion into the history of science.

---

[1] Some people claim that the stars on the moon should have had a different position in the sky than on Earth and astronomers should have noticed this if they had been seen. This is not correct in view of the fact that the stars in the sky are several billion times further away from Earth than the moon.

The Franciscan monk William of Occam lived from 1285 to 1347[2] and was not only far ahead of his time with his analytical way of thinking, but he also concerned himself with scientific subjects. Long before the Enlightenment, when the Ptolemaic world view of an Earth immobile at the center of the Universe was still valid, William of Occam thought about the nature of hypotheses. Through observations of nature and thought experiments, he implicitly came to the following conclusion: If a phenomenon can be explained by many different hypotheses or assumptions, the hypothesis that comes with the fewest free assumptions (parameters) is preferable, or rather, correct. This "principle of simplicity" is so far-reaching and successful that nowadays it forms the basis of scientific work, and it is widely known in the history of science as "Occam's Razor."[3] It states that explanations for natural phenomena should be valid with as few free assumptions as possible, and that you should always search for simpler explanations for the observed phenomenon. In other words, if I need fewer assumptions to explain something, it is a better explanation than one that has to make more assumptions. One of the conditions for applying Occam's Razor is the existence of several theories to explain the same phenomenon. It is certainly still possible that a newer theory is better, even if it is more complicated than the old one. A good example of this case is Einstein's theory of gravity, which is much more complex than Newton's theory. It has more assumptions but can explain a lot more observations.

The classic example of Occam's Razor in action comes from when the geocentric worldview of Ptolemy was replaced by the heliocentric worldview of Copernicus. Over the centuries, astronomical observations became more and more precise, which made the Ptolemaic model with the Earth at the center of both the universe and the planetary system inevitably more complicated, because it required an increasing number of independent assumptions to explain the observations. But the more accurate the measurements of the stars became, the similarities of the observations with the Copernican worldview of the Sun at the center of the planetary system became more obvious. The breakthrough came from the observations made by Tycho Brahe, on which Kepler's laws were based. Upon this foundation, Isaac Newton developed the Law of Gravity, which was formed with generally applicable equations instead of being tailored to a specific planet. The Ptolemaic worldview, however, became increasingly complicated in order to describe the more accurate

---

[2] It is exactly this William of Occam, by the way, that is the model for the novel character of William of Baskerville in the book *The Name of the Rose* by Umberto Eco.

[3] The principle is not directly found in Occam's writings. The term "Occam's razor" for the principle of economy was first formulated in the 19th century by mathematician William Rowan Hamilton.

observations until contradictions within the models became too obvious. The reduction of independent assumptions is therefore based on the demand for models that leave as little room for interpretation as possible. The aim is to explain the world consistently and logically rather than arbitrarily, so as to avoid making up fantasies that are not reasonably related to what is actually observed.

Turning back to the missing stars in the photos from the Moon, it quickly becomes clear that the assertion about technicians forgetting to hang them is problematic because it is too complex and has too many logically inexplicable consequences. First, you would have to explain how the studio managers could make such an easy mistake. An inconspicuous conspiracy should at least be clever enough as to keep millions of people from catching it. Perhaps the director intentionally wanted to send a sign of the fraud (he could have been under coercion from the Secret Service), but this in turn increases the number of free parameters in the sense of Occam's Razor. Secondly, the question arises as to how such a simple error could occur, even though a lot of money would have been available for the recordings and there would have been significant oversight. And thirdly, one may wonder how not even one member of the team (technicians, engineers, etc.) noticed this mistake. Last but not least, it must be pointed out that NASA still hasn't been able to solve the problem of stars being absent in their photographs, since there aren't any stars in any of their images from Earth's orbit. All of the stars are even missing from pictures taken during recent Space Shuttle and unmanned missions (Fig. 4.2). The only missions with photographs that include stars have explicit astronomical objectives or are those where pictures can only be captured at night (Fig. 4.3).[4] So, if the light bulbs were forgotten, you have to wonder how NASA could still be employing obviously incompetent technicians and yet, manages to get spacecraft and satellites into space. It shows that the simple assumption made by the Moon landing deniers has extensive consequences, all of which must be explained conclusively and in accordance with the established theory (forgotten stars on the studio ceiling). That all-encompassing explanation is very difficult to make.

For the purpose of utilizing Occam's Razor, we can introduce a new explanation and test whether it gets by with fewer assumptions. Luckily, we can perform the same experiment here as they did on the Moon, because there is not very much difference between the two places for the purposes of our

---

[4] Sensitive digital cameras were used for such sequences. Examples of films at night from Earth orbit, all with stars in the sky, can be found at http://eol.jsc.nasa.gov

Fig. 4.2 The *Hubble Space Telescope* photographed from the *Columbia Space Shuttle* STS-109 in March 2002. Photo: NASA. No.: STS109-730-034

experiment. Here on Earth, the stars are visible at night, and here too we have cameras to take pictures with—a completely analogous experimental setup.

QR: The International Space Station ISS flying over Africa. tinyurl.com/yajvldef

For this experiment, an old commercially available analog camera can be used to take pictures of the sky using different exposure times and with a fully opened aperture. With such a camera, a photographer typically exposes

Fig. 4.3 Night photo of the Southern Lights with stars above the horizon, taken from the International Space Station in September 2011. Photo: NASA. Product No.: ISS029-E-008433

moving objects (people, vehicles, animals) for about 1/60 of a second so they appear sharp and unblurred. Likewise, care must be taken not to overexpose the objects. By using a short exposure, the photographer also manages to keep their own restless movements from blurring the pictures (my own motor control, meanwhile, leaves a lot to be desired so the shorter I set the exposure, the better). However, if you choose such fast shutter speeds (short exposures) to photograph the sky, you won't see anything. No stars in the sky. If you now progressively increase the exposure time, you will discover that only with an exposure time of several seconds the brightest stars will finally become visible in the photo. This is an interesting result! Apparently, the stars in the sky are all so dim that they remain invisible to film for an exposure time of less than one second. This, of course, has consequences for our analysis. The exposure times on the Moon were chosen so that the image is not overexposed. The astronauts, the Lunar Module, the instruments, and the moonscape shine much more brightly than the stars (they are illuminated by the sun). Therefore, exposure times of several seconds are far too long because the objects would be washed out in white light. In addition, the moving astronauts should be photographed such that their image appears sharp. It's no different than on Earth. We are forced to choose a short exposure time so that the people in motion don't appear blurry. But with this short exposure time, the stars remain invisible at night. Nevertheless, everyone will be able to see for themselves that the stars have not disappeared.

If you now compare this experiment with the assertions of the Moon landing deniers using Occam's Razor, it is easy to see there is a clear winner. The number of necessary prerequisites for a complicated conspiracy in the studio is so immense and the plot has so many problematic consequences, that the explanation regarding camera exposure times is obviously favored. The only assumption in the latter was that the stars are not bright enough in relation to the scenery to be visible on pictures taken with the exposure times required for normal photography. In addition, we can prove this fact experimentally and without further assumptions, entirely in accordance with Occam's Razor and with the help of inductive reasoning (see Chapter 3). The inductive proof is as follows: At night you can't photograph stars with short exposure times. Therefore, it is impossible to photograph stars with short exposure times anywhere. This also applies to the human eye when confronted with very bright light sources. On the Moon, the astronauts could not see any stars with their eyes, because the sun outshined everything. This was confirmed by Armstrong and the other astronauts in interviews shortly after the Moon landings (Fig. 4.4).[5]

Neil Armstrong gave an informative interview to the BBC as early as 1970, describing that in the pitch-black, moonlit sky only the Earth can be seen, but no stars. This is an interesting statement at first glance, but of course, it was not an argument that there weren't any stars in the sky. Everyone can understand the explanation for this. Very bright floodlights are sufficient for clarifying this issue (e. g. floodlights on a football field). If you're looking for the stars past bright lights on a football field, you'll find that your eyes are strongly blinded and the relatively weak stars in the night sky are obscured. It's no wonder: the stars are about one billion times dimmer than the lights. If you compare this to the sun, which shines a hundred times stronger than lamps on a football field, stars being invisible in the night sky is inevitable. In the conversation, Armstrong also discusses his problems of estimating distances on the Moon. I'll come back to that in Chapter 7. His optimistic assessment of future lunar bases towards the end of the conversation is discussed in the last chapter.

---

[5] Most Moon landing deniers claim Armstrong never gave any interviews. But that is not true! On the internet, you can find a host of public appearances by Armstrong, including various interviews.

Fig. 4.4 A BBC Interview of Neil Armstrong from 1970. You can find the interview on the internet by using the search terms *Neil Armstrong BBC 1970*

Since the lunar missions were not astronomical missions (see Chapter 2), we found a simple and sound explanation for the missing stars in the sky. The argument of the Moon landing deniers that the lamps on the studio ceiling were forgotten is illogical and does not meet the requirements of a well-founded theory. Therefore, it cannot be used as an argument for a conspiracy.

5

But Look! The Flag Flutters!

Without air there can't be any wind. And without wind, a flag cannot flutter. Since there isn't any atmosphere on the Moon, you would expect that a flag wouldn't be moved by the wind. Planting a flag on the Moon was actually never thought about during mission planning. It was included so late in the program, that the procedure for planting the flag was the only activity for which the astronauts of *Apollo 11* did not train.[1] It turns out that Armstrong and Aldrin didn't manage to plant the flag firmly enough into the ground, so during their launch off the Moon in the Lunar Module the flag fell down.[2] But now critics point out that, in several NASA recordings, the American flag planted on the Moon flaps in the wind. Their argument is this: **In several video sequences, the flag flutters in the wind and therefore, the respective scenes must have been shot on Earth**.

Looking back at the issue of the stars missing from the sky, this argument is exciting in and of itself. We have seen in Chapter 4 that the missing stars are explained by someone simply forgetting to put the corresponding lamps on the ceiling of the studio where the scenes on the Moon were shot. So, the video would have been shot in a closed room—right? Then where did the wind come from? Were some scenes shot in a studio and others in the open air? In order to solve this contradiction, you would have to assume that the

---

[1] There were some initial thoughts to also plant flags from other nations. This idea, however, was quickly abandoned by Congress due to the argument that the mission was financed solely with US taxpayer's money.

[2] The political and technical aspects of the flag can be found at https://www.jsc.nasa.gov/history/flag/flag.htm

© Springer Nature Switzerland AG 2019
T. Eversberg, *The Moon Hoax?*, Science and Fiction,
https://doi.org/10.1007/978-3-030-05460-1_5

Fig. 5.1 The flag of *Apollo 11*. Photo: NASA/N. Armstrong. Product No.: AS11-40-5905

scene was shot in the studio, but a door was left open somewhere and the penetrating wind led to the flag's movement (some people suspect air conditioners are at fault). As with the subject of stars in the sky, here we would also have to ask ourselves how such a simple mistake can be made when planning a truly global fraud (Fig. 5.1).

But let's examine the assertion that the flag is fluttering in the wind a little more closely. In fact, there are a whole series of NASA videos in which the flag planted by the astronauts on the Moon moves. However, there is not a single sequence where the flag moves without one of the astronauts either touching the flag or its flagpole, or touching it a few seconds before it moves. A movement or "fluttering" can only be observed if the flag has just been or is being touched. The skeptics refer to different video clips in which the flag moves by itself. However, these clips are quite brief and they all start shortly before or directly with the corresponding flag movement. It leaves you to wonder what happened just before and after the clips. Fortunately, an extensive amount of video was taken on the Moon, and in many cases, it is evident that the selected short clips have a back-story. This back-story invariably includes astronauts handling the flag. On closer inspection you will find that complete sections of the video are withheld by the critics, and these sequences clearly show that the flag has been manipulated. One example of such a clip is a scene from *Apollo 14* where Alan Shepherd and Edgar Mitchell are being filmed by a remotely controlled camera as they plant the flag (Figs. 5.2 and 5.3).

Fig. 5.2 The commander of *Apollo 14*, Alan Shepard, and the US flag. Photo: NASA/E. Mitchell. Product No.: AS14-66-9232

Fig. 5.3 *Apollo 17* pilot Jack Schmitt and the US flag below the Earth. Photo: NASA/G. Cernan. Product No.: AS17-134-20384HR

QR: The astronauts from *Apollo 14* plant the flag. tinyurl.com/y7wkqbxr

You can see two things here: one is that the flag is not only attached to the vertical flag pole, but also to a horizontal crossbar. Therefore, only the lower free corner of the flag "flutters". Secondly, you can see that the flag only moves when the flag stand has been turned. The crossbar does not follow the flag as would be normal in the wind, but vice versa, the flag follows the pole in its movement. There is no gust of wind to speak of. This is exactly what the corresponding scene from *Apollo 17* shows; here, too, the flag follows the crossbar's movement and not vice versa. And once again, the flag only moves when an astronaut touches it.

QR: The astronauts from *Apollo 17* plant the flag. tinyurl.com/ybz6k55q

To be fair, I must admit that the corresponding pendulum motions actually look different than you would expect on Earth. In fact, you can easily get the impression that these are movements caused by the wind. This is especially the case in some scenes because the preceding touches from astronauts happened relatively long before the motion occurs. However, we must not forget that six times less gravity acts on the flag material on the Moon than on the Earth. By precisely analyzing the flag's motion, you can determine that an oscillatory response describes its movement without issue, quickly invalidating the assertion of flag movement caused by the wind.

Once again, I need to stress Occam's Razor. The claim that wind moves the flag requires an extraordinary number of assumptions and simply is not evident in the videos. We may quite rightly ask what is actually being "seen" here. If there is wind in the studio, then it always blows exactly when an astronaut touches the flag (or shortly afterwards, but never before), requiring an explanation as to where this wind comes from in a studio. Some critics believe that air conditioners or fans installed in the studio made the flag flutter—but then why doesn't the dust on the floor blow away?

Occam's Razor helps here again. We can make the recurring observation in the videos from the Moon that the flag only moves when touched and never moves without being touched. We also notice that when the flag is turned (this also only happens when touched), the crossbar to which the flag is attached always moves ahead of the flag and not vice versa, as would be expected if the wind blew on the flag and caused the motion. Therefore, not a single video clip proves the existence of any air currents. This is entirely in line with the findings from NASA archives, and Occam would have undoubtedly opted for this explanation.

6

A Lamp: Oblique Shadows

There is only one "lamp" on the Moon that can illuminate the landscape. This lamp also happens to be very far away—it is our Sun. It shines a light over the land on the Earth and the Moon, creating corresponding shadows. The Sun is so far away in relation to any setting on the Earth or Moon that we can actually treat it as if it were an infinite distance away. When we do this, we can assume that both the rays from the Sun and the shadows created by them run almost parallel to one another. Therefore, if you photograph the shadows, they should appear parallel in the picture.

However, there are many pictures from the Moon landings in which the shadows of the astronauts, devices, instruments, and rocks do not appear parallel at all. Not only that, but often even their lengths are different. There is a well-known photo that shows two astronauts standing next to each other and their shadows are neither parallel nor the same length. Anyone who looks at these pictures should be as irritated as I was when I first saw them, because this observation contradicts the fact that shadows created by a very distant light source should be parallel. This is exactly the objection raised by skeptics of the Moon landings. **If shadows that are caused by a relatively distant light source are not parallel or show significant differences in length, these shadows were created by additional sources of light**. In other words, since NASA never claimed that additional lights were taken to the Moon, the Moon landing scenes must have been lit with multiple lamps in a studio.

This argument does not raise any objections; rather, it is entirely correct—shadows that are created by a single, far away light source must run parallel to one another. But as it pertains to the photos taken on the Moon, the conclusion is wrong! The claim that all shadows caused by a distant light should be

© Springer Nature Switzerland AG 2019
T. Eversberg, *The Moon Hoax?*, Science and Fiction,
https://doi.org/10.1007/978-3-030-05460-1_6

Fig. 6.1 Buzz Aldrin, pilot of *Apollo 11*, next to the Lunar Module. Photo: NASA/N. Armstrong. Product No.: AS11-40-5902

parallel, i.e. *depicted* in the same direction, is not correct. The little "trap" that conspiracy theorists set, and that I deliberately set for you, can be found in the word "depicted". In the case described above, it is absolutely correct that all shadows must run parallel, but not that they are *depicted* in parallel (Fig. 6.1).

To find the cause of this completely human error, we need to consider how humans perceive the world and what exactly happens in the process. The process of seeing is so commonplace that you rarely realize what you are actually doing. You have two eyes and live in a world that appears three-dimensional. With these two eyes you can distinguish between moments about 20 milliseconds apart, and thus you perceive everything around you at a rate of about 50 frames per second. In addition, you make a "trigonometric" measurement with these two eyes with which you can perceive depth. A one-eyed person would immediately understand what I'm talking about. When you look at a nearby object, you can only estimate its distance from you or its relative position if you look at the object from at least two different positions. These two positions are afforded by your two eyes, or you can move your head. Just give it a try. Close both of your eyes and ask another person to position an unknown object (we want to exclude everyday experiences) in front of you on a table. Now open only one eye and try to estimate how far away the object is. You will see that this is difficult without the second eye. This is something that one-eyed people experience every day. But if the object in question is not nearby but rather a great distance away, the separation between your eyes is no longer sufficient to count as viewing something from two separate points. Instead, you must look at the object while standing at two different positions. Another way of estimating the distance to something is to move towards or away from the object that you're looking at. The resulting resizing of your view of the object as it becomes closer or farther away gives you an idea of the size of the object. We each do all of that around 50 times per second, and the process is so natural that we rarely give it any thought.

Returning to our problem with non-parallel shadows, the issue is that we are always "one-eyed" while viewing normal photographs.[1] The camera uses a single lens, and the corresponding photos it takes cannot convey the three-dimensional information that our eyes typically use. This hinders our reliable assessment of the landscape. Information about the texture and layout of the ground is incomplete, and this plays an important role in evaluating the length of shadows, which I'll illustrate with an example.

If you look at the length of your shadow at sunset, when the sun is at the height of your eyes, you will notice that your shadow is significantly longer than the actual size of your body size. The shadow may stretch out across the ground many times your height. This is obviously a projection effect on a very tilted surface, in this case the ground. However, if you measure your shadow in front of a wall whose surface is perpendicular to the direction of the sun, i.e. a vertical wall, your shadow will be about tall as your body. From this, you can see quite clearly that the slope of the surface on which the shadow is projected has an influence on the length of the shadow. If two people stand next to each other, and the inclination of the two surfaces on which the individual shadows are cast is not identical, the shadows will appear to be different lengths. This is very often the case on uneven ground, like on the Moon. On such a surface,

---

[1] I'm neglecting here 3D images and their special technology.

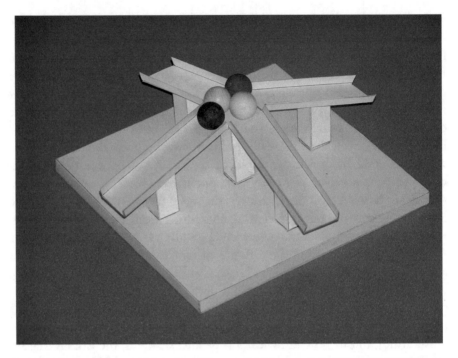

Fig. 6.2 Four ramps on which balls roll "upwards"! Source: K. Sugihara/Meiji University, Japan. The next QR code gives the corresponding clip

deviating shadow lengths are not all that unusual, but rather completely normal. If we do not receive any additional three-dimensional information from a camera, i.e. we cannot evaluate the surface topology, then the different shadow lengths are not surprising and can be explained with ease (Fig. 6.2).

The shadows cast in different directions in the photos are projection effects from mapping the three-dimensional world onto the two-dimensional representation of film. This has important consequences for our perception of the geometry of the image. While it may seem unusual at first, an example makes this immediately clear as you can understand the phenomenon at home without much effort. Take a picture of the tiles in your bathroom. It is clear to see that the tiles are more are less parallel (the "more or less" depends on how good your tile layer was). However, when you look at the photograph, depending on the angle at which the picture was taken, the lines of the tile joints run radially outwards in different directions—not appearing by any means parallel. If you don't have tiles in your bathroom, you can try another method. On a sunny day, take two pencils and stick them into the ground a few centimeters away from each other. You'll take pictures of the two pencils with your camera from a few meters away, but first tilt one of the pencils slightly in the direction towards where you'll take the picture from and tilt the other pencil slightly away in the exact opposite direction. If the Sun illuminates these two pencils from a different direction than the direction you take the picture from, the shadows cast by the pencils will clearly not be parallel. However, because you tilted the pencils only in the directions towards or away from the camera, the pencils will appear parallel in your picture.

QR: Balls rolling uphill. tinyurl.com/yac7zgl7

Thus, you can undoubtedly see from these two examples that pictures or even videos never depict reality, but that we interpret these things with our life experiences. This interpretation leads to the fact that we usually think non-parallel shadows in pictures are completely normal. Painters know this fact well. Try to paint a picture of railway tracks heading directly away from you towards the horizon, without worrying about your artistic ability. You will notice that to provide the effect of depth you have to draw the rails in the picture converging from bottom to top, and certainly not parallel (for your own safety, just use your imagination and don't stand on real train tracks!). If you're like me and don't have much talent for painting, you'll quickly notice how difficult it is to depict depth on canvas.

But there is still another fact that is always overlooked! If someone claims the shadows that are cast in different directions are made by two or more spot-lights, then they would naturally have to explain why two objects or astronauts standing next to each other do not each cast two or more individual shadows, running in different directions. Try putting two lamps next to each other in your home and check the shadow cast by a single object. You will notice there are two shadows cast by the one item. Every spectator of an evening football game under floodlights has seen this effect. So then why don't the astronauts and objects on the Moon cast two or more shadows simultaneously in any given photo, even though their individual shadows are not parallel?

Oblique shadows and their unequal lengths are an effect of missing three-dimensional information. Even further, we humans take the complexity of our process of perception for granted because we use it every day (actually, every second), and we barely distinguish between using it for viewing the real world and illustrations any more. This is important in the case of evaluating unusual landscapes for which we have little basis for comparison. If this isn't

Fig. 6.3 Image of the rock collection tool from *Apollo 16*. Photo: NASA/J. Young. No.: AS16-108-17697

taken into account, then there truly is no explanation for the oblique shadows other than having additional lighting. Everyone interprets pictures and videos as two-dimensional representations of the real world. It would be difficult to manage everyday life in a three-dimensional world otherwise. Surprisingly, the critics of the Moon landings do not think about this when considering the pictures from the Moon landings. However, since the camera and photographic conditions on the Moon are basically no different than those on Earth, the way we perceive the images should remain the same. You should therefore be skeptical of arguments related to an inconsistency between the shadows. It also isn't a bad idea to pay attention to shadows throughout your day every once in a while (Fig. 6.3).

7

Manipulation of the Pictures

One of the most popular means of discrediting the Moon landings is claiming that the pictures were photoshopped, i.e. the presented photographs were manipulated. Some of the more spectacular examples include the photo of three astronauts on the Moon (everyone knows that there were always only two on the Moon) or photos including items that were never actually taken (sales signs, camels, and even whole ships—a camel in the lander, marvelous!). I think that we can classify these manipulations as "jokes" and "advertising," and as such, they do not require our consideration because they are so easy to catch and understand. Rather, I'll focus on original photographs that may appear questionable to the viewer and that are used as "proof" of the Moon landing conspiracy.

Let's start with the calibration marks from the cameras. NASA had sourced a series of cameras from a company called Hasselblad that were designed specifically for use on the Moon. The cameras had parallel calibration marks or lines in their focal plane, so that the marks would show up in the images as dark, sharp, and parallel with identical distances between one another. They were used in image analysis to measure distances on the Moon, because it was expected that distances and proportions would be difficult to determine without known calibration marks (see Chapter 6). Therefore, all original photos show these calibration marks as a grid network.[1]

Interestingly, upon closer inspection there are a few pictures in which something seems wrong. The calibration marks, which were inserted in the

---

[1] Incidentally, the cameras did not have a viewfinder shaft and the astronauts were unable to target objects.

© Springer Nature Switzerland AG 2019
T. Eversberg, *The Moon Hoax?*, Science and Fiction,
https://doi.org/10.1007/978-3-030-05460-1_7

Fig. 7.1 Charles Duke on the rover of *Apollo 16*. The zoomed in section shows the S-band antenna, which "covers" a calibration mark. Photo: NASA/J. Young. Product No.: AS16-107-17446

focal plane of the film, seem to be covered by bright objects in the photo (Fig. 7.1). In one case, a calibration line is interrupted by a white stripe in the US flag. In another case, various white antennas appear to cover the calibration lines. That's supposed to be impossible. How can a calibration mark be displayed in the focal plane of the camera behind (!) the photographed object? Hence the argument: **Calibration marks inside the camera cannot be covered by objects in the background, therefore the images were manipulated**.

From a review of the facts, this appears problematic at first glance. How is it possible to corroborate photographs taken several decades ago somewhere around 300,000 kilometers (186,000 miles) from Earth? The answer lies in performing another experiment on Earth, just as we did in Chapter 4. Unlike the Moon, we have an atmosphere here on Earth, but otherwise there are identical conditions for taking photographs. Therefore, we can take what we learn from taking pictures here on Earth and assume that similar effects will be seen in pictures taken on the Moon. Just like the astronauts did on the Moon, you can photograph very bright objects. After doing this, you will quickly realize that light surfaces tend to outshine more dim areas around the edges. It is much more difficult to capture a picture of a thin object, such as a thread, against a light background than when it is placed against a darker background. And if the thread is placed in front of a pattern of light and dark stripes, the thread will appear to be disrupted in front of the lighter stripes. The bright background outshines the thread. This is not a manipulation of the picture, but rather it is the effect of strong contrast. By the way, this is analogous to human perception with the eye. If you look into a strong spotlight, it is almost impossible to distinguish fine structures.[2]

Another argument: **The calibration marks also disappear on photographs that do not depict any brightly shining surfaces**. Images from the internet are cited as "proof" of this. I've gotten used to checking the digital image format as a first order of business. In almost all cases the images are in the so-called JPEG format. This standard format, developed by the Joint Photographic Experts Group (JPEG), describes an image compression process with which images containing a significant amount of data are made usable for the internet. While the compression method is adjustable, this format has the disadvantage that fine details may be lost in the process depending on the available storage space within the compressed image. This is generally the case with the considerable compression used for images uploaded to the internet. If you want to examine high-resolution images of the missions, you should get the images from the web archives of NASA and not the compressed versions simply to speed up the downloads (Fig. 7.2). Keep in mind that even the high-resolution images have already suffered a reduction in image quality due to the scanning process of the original analog photographs. The history of an image is also important. If the history is unknown, or rather, if what the previous owner of the images may have done with them before is unknown, you should exercise caution in evaluating the images and coming to any conclusions.

---

[2] The contrast issue is also a fundamental problem in the search for extrasolar planets in astronomy. The extremely bright star outshines its entire neighborhood so that extremely faint planets literally are covered up by the light.

Fig. 7.2 The original high-resolution version of the previous photo. The zoomed in section again shows the S-band antenna, which "covers" a calibration mark. Photo: NASA/J. Young. Product No.: AS16-107-17446

A completely different analysis is needed for the series of photos that have different foreground scenes, but apparently identical backgrounds. At first, this doesn't seem to be very special because astronauts can actively get out of the way of any picture—no problem. But if the whole lander is shown in several shots and the lander has disappeared in a different picture with the exact same background and proportions, then there seems to be a problem. A well-known example is a picture of the Lunar Module of *Apollo 17* in the foreground with some hills in the background. A second picture shows the exact same landscape, but now without the lander. Neither the hills, nor the perspective appear to have changed, but the lander is suddenly missing. How could that be? Where could the lander have gone, even though nothing else

Fig. 7.3 Two pictures of the Taurus-Littrow Mountains of *Apollo 17*, with and without the lander. Photos: NASA/G. Cernan. Product No.: AS17-134-20508 and AS17-140-21500

changed in the picture? How can this same landscape be photographed once with and once without the lander? **Obviously, critical studio recordings were accidently released before the lander could be pushed into the scene** (Fig. 7.3).

To take a closer look at this scenario, I'm going back to Earth. This time, however, I will take into account the differences to the Moon. If you ever go to Cologne in Germany, you will notice that the people who live there are proud of their medieval cathedral, the Dom. On its southern tower, at a height of 97 meters (318 feet), there is a viewing platform from which you can enjoy a wonderful view of the surrounding city.[3] On clear days, you can see the Siebengebirge (Seven Mountains) far to the south near Bonn, and a popular game is to guess how far away the mountain range is. The estimates usually aren't that bad (it is just under 35 kilometers, or 22 miles away). There are a few reasons the estimates are pretty decent. The haze in the air caused by clouding around the mountains intuitively tells us that they must be a good distance away. We can also use objects in the foreground (e.g. buildings) to give us a frame of reference for size comparison. In the absence of any haze and any objects that we can use for comparison, it becomes very difficult to estimate distance. If you've ever looked at any impressive pictures of the Alps

---

[3] As it so happens, the command capsule was also located at this elevation on the Saturn V Moon rocket. Once you have stood on the south tower of the Dom, you can get a feel for the enormous size of that machine.

Fig. 7.4 Picture taken by geologist Jack Schmitt at "Tracy's Rock" and an enlargement of the right field of view. Photo: NASA/G. Cernan. Product No.: AS17-140-21496

where neither people, houses, nor clues are depicted, and suddenly you find out that the mountains pictured are not the Alps, but in fact are photos of the Himalayas instead, you can understand this deception. Photographs make a good distance estimate particularly difficult, as we also have no information about the lenses used and their focal lengths,[4] in addition to only being able to look with "one eye" in relation to distance, as explained in the previous chapter. For example, the Sun or the Moon often appear much larger on the horizon than they are in reality. The movie poster for "E.T. the Extra-Terrestrial" is a great illustration of this: the child hero is shown with his bicycle flying in front of a huge Moon. Anyone can take pictures of such an enormous Moon on the horizon by using a telephoto lens with a long focal length.

All of this happens in the pictures of the Moon lander from *Apollo 17*. In reality, the lander wasn't moved, but it can be shown with the help of other pictures from the mission (they are available to everyone on NASA servers), that it is about two miles away in the photo where it isn't shown (Figs. 7.4 and 7.5). The reason why the background has not changed is simply that the hills in the background are actually the Taurus-Littrow Mountains with peaks up to 3000 meters high (9842 feet) and several dozen miles away. A foreign environment with no basis for comparison makes it very difficult to estimate distances correctly. If, on top of that, there isn't any atmosphere and the

---

[4] For the Hasselblad cameras on the Moon, the company ZEISS developed the *Biogon 5.6/60 mm* lens with very high contrast and sharpness with the maximum freedom from distortion.

Fig. 7.5 Telephoto view of the enlargement of AS17-1140-21496 with the Moon lander at a distance of one and a half miles. Photo: NASA/G. Cernan. Product No.: AS17-139-21204

settings of the camera are also unknown, reliable estimates of distance are virtually impossible.

To be able to adequately evaluate images from the Moon landings, it is necessary to look at more than just individual images. You have to analyze all of them in their collective context. This includes the history of the images, the technology used to capture them, and the ambient conditions. If you do not consider all of these things, you can easily be misled.

8

Is Everything in Slow Motion?

The Moon is significantly smaller than the Earth and as such, the gravitational force is lower. Everything on the Moon weighs about six times less than it would on Earth. While the resulting slow-motion-like movements should be no surprise, they are immediately noticeable in film and television recordings. The physical fact of lower gravity is accepted by everyone, even by people who doubt the lunar landings. However, skeptics throw in this: **All of the videos were recorded in studios on Earth and were shown to the public in slow motion to simulate a lower gravity**.

This trick was depicted in the Hollywood science-fiction movie *Capricorn One* from 1978 in a thrilling and very suggestive way, and the thought that NASA could have done just that is absolutely understandable. This doesn't seem implausible if you consider the daily use of slow motion in various sports broadcasts on television where you can watch people fly through the air amazingly slow. However, very few people actually wonder whether it is at all possible to simulate the lower gravity on the Moon by using slow motion playback. At least there are a good number of video recordings available to answer this question. These recordings that are supposedly played in slow motion do not only show the simple bouncing motion of the astronauts, but also complex motor functions, all of which would have to be simulated in their entirety. The original recordings thus offer an extensive pool of information that is well suited for checking the above assertion. One might object that the counterproof should be conducted independent of NASA's data. But the exact opposite is the case: NASA's recordings are being called into question and that's why those recordings are what should be considered (Fig. 8.1).

© Springer Nature Switzerland AG 2019
T. Eversberg, *The Moon Hoax?*, Science and Fiction,
https://doi.org/10.1007/978-3-030-05460-1_8

Fig. 8.1 John Young jumps up in front of the flag while saluting. Photo: NASA/C. Duke. Product No.: AS16-113-18339

QR: John Young's jump. tinyurl.com/ya9qwffg

The perfect clip for closer inspection is a sequence from *Apollo 16* where astronaut John Young is standing next to the US flag in front of the Lunar Module being filmed by his co-pilot, Charles Duke. Brimming with patriotism, he asks Young to pose while saluting and standing next to the flag. John Young salutes, but at the same time he jumps up and lands on his feet again (the flag did not move in this sequence, by the way). The height and duration of the jump are what we're really interested in here. The astronaut's jump is, in fact, a simple physical experiment in a gravitational field. Students may remember this experiment as "The Vertical Throw." This experiment is useful because it shows the relationship between the governing parameters: gravity, jump height, and duration of the jump. Isaac Newton was the first to successfully describe this relationship in 1687 with the publication of his law of universal gravitation. Because this law is not only well known, but also one of the foundations of modern science, I presume that it will not be called into question by anyone who doubts the lunar landings.

First, let's check whether the same type of jump would even be possible on Earth. Knowing Young's height, you can estimate that he jumped about 44 centimeters (17 inches) high. This alone would be an extraordinary achievement on Earth when jumping from sandy ground, without making a running start, and without squatting down. Well-trained ski jumpers can jump from a squat to a height of around 50 centimeters (19.7 inches), and the personal record for the famous ski jumper Sven Hannawald for such a jump is 51.6 centimeters (20.3 inches). This jump height is only surpassed by high jumpers and dancers with a full-strength attempt. Therefore, you can assume that jumping that high while barely using your knees, like Young's jump, is completely impossible on Earth—try it yourself. This is especially true if you are carrying a life support pack on your back. Many critics of NASA now argue that the backpack wasn't as heavy as they claim; instead it is just a hollow frame made of lightweight metal with a fabric cover (a life support backpack is unnecessary in a studio). But the problem remains that even with a weightless backpack, combined with an equally weightless but bulky space suit, a 44-centimeter jump from sandy ground would be extremely implausible. One might object that for the astronauts, who were well-trained pilots, such a jump would actually be quite possible. Or even that, in reality, the people in the suits weren't astronauts, but athletes. Still, this argument isn't very strong if you consider Sven Hannawald's abilities.

Going back and forth on how high a person can jump is only of limited help in getting to the truth of the matter, so for simplicity's sake, let's assume that an athlete wearing gym clothes and not carrying a life support backpack can jump to a height of 44 centimeters (with or without a running start is irrelevant here). Let's now investigate how much slower the film would have to be unwound in order to simulate the lunar videos. In order to do this, we have to analyze the entire jump, not just the jump height. In other words, we have to consider how long it took to complete the jump. Using simple physical observations (keyword: "The Vertical Throw") and some calculations, you can figure out that the jump on Earth should take exactly 0.6 seconds from takeoff to landing. In the film, however, the whole sequence takes 1.47 seconds. This means that the film must be unwound slower by a factor of 1.47 divided by 0.6, i.e. by a factor of 2.45.

This is an important result because some skeptics claim that the movements of the astronauts would look completely natural if the videos were played at twice the speed. But we've just discovered that it would take 2.45 times the speed of the playback instead of 2 times the speed to make the jump on the Moon look like a regular jump on Earth. At first glance, the difference from the expected result that we see in the jumping sequence seems inconsequential and almost petty. However, it is a first piece of evidence that the Moon landings were real. The number 2.45 is the square root of 6. In the physical relationship between jump height and acceleration due to gravity, the duration of the jump is squared.[1] Since the acceleration due to gravity on the Moon is only one-sixth of that on Earth, the result of the root of 6, i.e. 2.45, is fundamentally necessary; and luckily, that is exactly the number that we arrived upon from analyzing the video of the jump. John Young and Charles Duke unwittingly made an independent verification of the Moon's mass, through determining the strength of the Moon's gravitational field, by joking around. It is unclear why the factor of 2 is often mentioned by skeptics, but it is clearly wrong. This film sequence therefore offers no evidence against the Moon landings. And to ask a further question: If this scene was played in slow motion, how could the engineers be so clever as to choose the physically correct slow-motion speed, but so careless that they forgot to put the stars in the sky of the studio?

Now that we've determined the proper speed that a video shot on Earth would have to be played back at to appear as if the jump were made on the Moon, we can subject all other recordings to this test. Let's take videos from the Moon and play them back 2.45 times as fast. If the videos were shot on Earth, the astronauts' movements should now look realistic. But if you do that, you'll see some very peculiar things. If all the videos were played back at an accelerated speed, then not only should the movements affected by gravity now appear to occur at a normal speed, but all other movements should also appear to occur at a normal speed. For example, this would include arm movements. In the film sequence in question, Young salutes next to the flag.

---

[1] The distance s ($s = 2 \cdot h$, with the jump height = h) is calculated as $s = \frac{1}{2} \cdot a \cdot t^2$ with the acceleration $a = 9.81 \ m/s^2$ and the jump duration = t. Solving for time t results in $t = \sqrt{2 \cdot s / a}$. But because the acceleration on the Moon is six times lower than on Earth, the ratio of the jump duration between Earth and the Moon will be the square root of 6, or 2.45.

But played back at an accelerated speed, this salute gets the slapstick effect characteristic of early cinema because the movement of the arm is too fast now. Young moved his arm normally, as you can see in the original recordings without accelerated playback. This contradiction is a clear indication that the films were not played back slower originally, and that the jumps actually took place on the Moon. In addition, there is still the issue of him jumping 44 centimeters high, which a top athlete can only accomplish in its same form with difficulty. Aside from that, with the jump time taking 1.47 seconds, the astronaut would have to jump 2.64 meters (8.7 feet) on Earth in order to make the jump last that long, thus breaking the world record for the high jump.

Other video clips raise more questions: how could the astronauts effortlessly get up from a kneeling position without using their hands for support? Or how is it possible to get up from a push-up position without some momentum? During their second trip to the lunar surface, John Young falls to the ground while carrying penetrometer for measuring soil density. Afterwards, he swings himself back up again in a way that would be completely impossible on Earth without aids. Even if this clip is played back faster at the correct factor of 2.45-times speed, it now appears even less convincing. These shots are a clear indication that the films are by no means played back in slow motion. These movements would be impossible with the special effects of the 1960s or even with ropes. Further still, the whirling up of dust can't be explained without the lower gravity of the Moon.

QR: John Young's fall. tinyurl.com/ydx26vpe

QR: John Young's fall at 2.45 times speed. tinyurl.com/ycdpwel6

Therefore, the slow-motion playback argument is in fact one in favor of the Moon missions and not against them. Young and Duke's physical experiment proves that only one-sixth of the Earth's gravitational pull prevails on the Moon, and they have thus confirmed a well-known observation of physics as explained by Newton's laws of gravitation. The claims of slow motion play-back, however, cannot be substantiated by observable facts. They are also inconsistent with other claims from a logical point of view. For these reasons, such claims are not suitable for proving that the Moon landings were faked.

9

Telescopes Can See Everything

Out of necessity, the lunar missions were carried out very efficiently. Any dead weight was quickly removed to keep the Moon rocket within the operational limits of its design; thus, defining the principle of rocket staging.[1] Once a stage's fuel tanks were emptied, they were dropped or left behind (rockets are really nothing more than huge fuel tanks with an engine connected at the bottom). This method is still used today by all space-faring nations and, for technical and financial reasons, there are no realistic alternatives to it. The situation is the exact same on the Moon: the descent stage of the Lunar Module, including its engine and tanks, was left behind as it was used as a launch pad for the rest of the vehicle's return to Earth, and remains there today on the surface of the Moon. The descent stage with the four projecting spider legs has a diagonal diameter of 9 meters (29.5 feet). The argument is now this: **Humankind has used their telescopes to observe galaxies millions of lightyears away; therefore, the lunar descent stage should be easily observable on the Moon with a telescope. However, not a single astronomer has been able to provide a picture of it so far. Consequently, the descent stage must not actually be on the Moon, and the whole story is a lie** (Fig. 9.1).

This argument is astonishingly clear and easy to understand—yet it doesn't only ignore the physics of optical systems, but also our everyday experience. Let's consider a counter-question: "How can I see structures on the Moon 300,000 kilometers (186,000 miles) away so easily and still not be able to see

---

[1] Without the multi-stage principle, the Moon rocket would have needed to be the size of the Empire State Building and would not have flown at least until the end of the 1960s.

© Springer Nature Switzerland AG 2019
T. Eversberg, *The Moon Hoax?*, Science and Fiction,
https://doi.org/10.1007/978-3-030-05460-1_9

Fig. 9.1 The four telescopes of the European Southern Observatory (ESO) on Cerro Paranal (a mountain in Chile) are among the largest telescopes in the world. Photo: ESO/H.H. Heyer

a friend's face just a few hundred meters away?" The answer is obvious: "The structures of the Moon are so much bigger!" There you go, it's that simple! You can recognize the absurdity of this argument effortlessly, using only your intuition.

To understand the connection on a fundamental level, you have to think about how the actual size of an object you can see is perceived and determined. Then you need to consider what the limits of perception are. To do this, let's look to the physics of light and its propagation. Don't worry, I'm not talking about the complex wave mechanics of photons, which are also difficult for physicists to understand—we are only interested in the geometry of light propagation. If you want to look at objects with a large-scale (macroscopic) angular size in the sky, e.g. the Moon or galaxies, you can use classical geometry optics. Trigonometric measurements and theorems (recall the sine, cosine, tangent, etc. theorems) are the basis for classical geometry, just as you may remember from your high school math class. If, on the other hand, you want to investigate miniscule angular sizes, asking how small an object can be for you to still be able to recognize the details of it (e.g. a friend's face nearby or the Lunar Module on the Moon), then you have to determine its optical

Fig. 9.2 The spiral galaxy Messier 51 in *Canes Venatici*, taken by the Hubble Space Telescope. Photo: Hubble/ESA—Hubble Space Telescope/No. heic0506a

resolution. But what you can optically resolve depends on the tools you have at your disposal. Obviously, using a telescope can help you to see details from a much greater distance away than if you only looked with your eyes (Fig. 9.2).

1. **Angular Size**—If you look at an object standing at some distance away, you won't get any information about the geometric size of the object, in other words, about its true size. Let's consider a car 1000 meters away. Naturally, you know how big a car is, but only from your own direct experience. From the perspective of perception, it doesn't matter whether it is a real car 1000 meters away or if it is a toy car that is 100 times smaller and 100 times closer (10 meters away). With the correct choice of distance and true size, both cars will appear to be the same size. Try it yourself by holding a stick in your outstretched hand in front of your face with one eye closed. With your other hand, hold a second stick that is half as long as the first stick in between your eye and the first stick. Adjust the position of the second stick until it completely obscures the first stick behind it. When you've accomplished this, you'll find that the distance from your eye to the shorter, second stick is exactly half as far as the distance from your eye to the longer, first stick. Even though the two sticks are different lengths, they

appear to have the same length when placed at the proper distances. With this simple experiment it becomes clear that the true size of an observed object can only be determined if you know the angular size of it *and* the distance to it.[2]

2. **Optical resolution**—Any optics, whether your eyes or a large telescope, can recognize two adjacent objects as two separate objects if they are far enough apart. Again, the angular size is the important measurement, but this time it is about the minimum angular distance between two objects at which the eye or the telescope can just barely recognize the objects as two separate objects. This so-called "resolving power" is an effect of the wave-like nature of light and it is determined solely by the diameter of the pupil or telescope optics and nothing else.

Let's look at the human eye. Its pupil diameter automatically adjusts to the brightness. If you look at a bright light, your pupil quickly shrinks in diameter; otherwise, you would be blinded immediately. In the dark, however, it opens much more slowly and only after a few minutes, the pupil reaches its maximum size—usually about 6 millimeters (0.25 inches). This enables you to perceive two separate objects that are at least 10 centimeters (4 inches) apart at a distance of 100 meters (328 feet) away. Modern large telescopes are up to 2000 times larger than the pupil of the eye and can, to stick to our example, optically resolve objects at a distance of 100 meters away that are 50 micrometers (0.002 inches) apart. That's the thickness of a single strand of hair—amazing! The key point here is that the optical resolution must always be expressed as an angle (or a distance away and a width or diameter of the object). This optical resolution remains fixed for the same optics system (i.e. your eyes or a telescope). A statement about the spatial separation of the objects must always refer to the distance the objects are away from the viewer, and thus the angular size comes into play. At a distance of 100 meters, my eyes can resolve two separate objects if they are at least 10 centimeters apart. But at a distance of 1000 meters, that minimum separation distance becomes 100 centimeters (25 inches).

It quickly becomes clear that the world's largest telescopes can see galaxies at distances of several hundred million light-years away, but they only have a spatial resolution of several thousand light-years wide at the distance of the galaxies. This is so huge that it is impossible to distinguish individual stars in a cluster of closely grouped stars. It is completely hopeless if you want to

---

[2] With this small experiment, we have used a set of theorems from elementary geometry.

resolve the planets of these star systems in distant galaxies, because the diameters of the planets are several million times (!) smaller than the optical resolution of the telescope at those distances.

If we apply this unavoidable fact (nature cannot be cheated) to the Moon, this means we get a resolution of about 20 meters (66 feet). So even with the largest telescopes in the world, we can't resolve the Lunar Module and any other equipment left on the Moon. Even with a 40-meter diameter telescope, as currently being planned, the Moon lander would only be seen as a blurry spot and we would still have to deal with the accusation that the new photos don't provide any evidence of the lunar landings.

The argument that it should be easy to see an object on the Moon with a telescope if it is possible to see distant galaxies with one has no physical truth. It's wrong!

10

Warning! Hazardous Radiation!

The universe is not a friendly place. You must bring your own air to breathe and you should protect yourself from the high-speed micrometeorites flying around all the time with a robust shelter. Both problems can be solved to some extent with space capsules or stations, which nowadays offer pleasant, albeit expensive, accommodations (palatable food, bathing facilities, toilets, and private sleeping quarters).[1] Although a hotel bar still needs to be sent up, these should make your stay bearable for a short duration.

But there is still another threat of a completely different caliber, one you can neither see, feel, nor taste: radiation. The universe is teeming with high-energy radiation and particles whose still partially unclear origins lie somewhere in the depths of the cosmos. The effect of radiation on biological organisms, however, has been well classified. Exposure to high-energy radiation for too long can cause serious damage to your body and could even lead to death. The strongest source of radiation for us is the Sun, whose sporadic eruptions are carefully observed by NASA during space missions. The progress of the solar eruptions, either solar particle events or coronal mass ejections, is closely monitored as they approach Earth (Fig. 10.1).

The activity of these solar outbursts fluctuates cyclically with a period of about eleven years. As it happens, all the lunar flights were performed during the peak of that cycle with the most frequent and strongest outbursts—it turns out that President Kennedy was not a solar physicist. The probability of

---

[1] At the time of the Moon landings there were no toilets onboard. The human necessities were adventurously dealt with by using plastic bags.

© Springer Nature Switzerland AG 2019
T. Eversberg, *The Moon Hoax?*, Science and Fiction,
https://doi.org/10.1007/978-3-030-05460-1_10

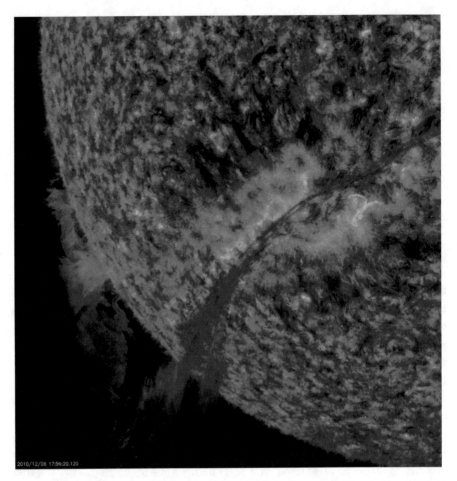

Fig. 10.1 Photo of a solar eruption, taken by the Solar Dynamics Observatory of NASA on 6.12.2010. Photo: NASA/SDO

a strong eruption, and the risk it would have posed to the crew, was very high during the flights to the Moon. NASA openly admits that all of the lunar astronauts were seriously risking their health and their lives. Such high stakes apply to most pioneering expeditions. The group of explorers with Roald Amundsen all put themselves in mortal danger on their way to the South Pole, just as Edmund Hillary and Tenzing Norgay did on their way to the summit of Mount Everest.

A particularly unpleasant area for humans is within Earth's radiation belts (also called the Van Allen Belts, named after the man who discovered them). These two radiation belts surround the entire Earth, one between the altitudes of 700 and 6000 kilometers (435 and 3700 miles) and the other between

15,000 and 25,000 kilometers (9300 and 15,500 miles). They protect Earth's inhabitants from the dangerous solar outbursts, and without them, there probably wouldn't be any life on Earth. The belts are generated by the Earth's magnetic field. They contain a significant concentration of high-energy particles, such that a prolonged stay in the belts would have serious health consequences. All of this is true even when the solar activity cycle is at a minimum. And we already know how radioactive radiation affects the human body due to the atomic bombs dropped on Japan and from later nuclear radiation tests.

If you want to fly from the Earth to the Moon, then passing through the Van Allen Belts is inevitable. Nonetheless, you should still try to avoid a high dose of radiation, if possible. The argument posed is this: **Because of the strong radiation in the Van Allen Belts, all lunar astronauts would have been exposed to lethal doses of radiation. A flight to the Moon is, therefore, practically impossible**.

Natural radiation also occurs on Earth. It can be emitted by geological sources such as radon gas and potassium. The amount of energy deposited in the body and the associated harmful effect depends on the dose (intensity of the energy) and the duration of exposure to this radiation. Since natural sources of radiation only have low intensities, people can live around them without any issues. This is different in areas where high radiation levels occur, for example, certain areas of the nuclear power industry, in radiation medicine, or in the Van Allen Belts. In such areas, you should take care not to be exposed to the radiation for too long.

I keep hearing that thick lead walls are necessary to effectively protect against radiation. Understandably, this is what most people imagine must be done, because they have no knowledge of nuclear physics. The solar wind essentially consists of helium nuclei, protons, and electrons. In contrast to high energy electromagnetic waves, all particles have very low penetration depths when they come into contact with any material. Consequently, using thin layers of material can keep the effects of heavy particle radiation within acceptable limits, as was planned for an emergency scenario during the Apollo missions. In the event of a solar storm, the thickest-walled areas of the space capsules or the fuel tanks would have been turned in the direction of the incoming solar wind. Lead shielding would only be useful when dealing with harder electromagnetic radiation, so-called gamma radiation, which has a very high penetration depth. A typical example of electromagnetic radiation is the X-Ray that you get at the doctor's office. But only a negligible amount of hard gamma radiation comes from the Sun, and therefore it is unnecessary to protect against such radiation during space flights.

While the strength of radiation is expressed in various units, the dose equivalent "Sievert" takes into account the biological effect of the radiation and is used for understanding doses in humans. Typically used in medicine, it is well suited for evaluating radiation and its harmful effects. Unfortunately, the units typically used in nuclear physics are unintuitive to use because we do not have a relative basis for perceiving them. Nevertheless, you should always be aware that strong radiation can have fatal effects and that the effect of radiation accumulates in the body over time.[2]

But let's get back to the claim that you receive a deadly dose of radiation in the Van Allen Belts, and look at a few comparative values:

- 0.5 millisieverts (0.0005 Sv) is the maximum annual radiation dose from space.
- 2 millisieverts (0.002 Sv) are typically received per year from artificial radiation sources (e.g. X-rays). The maximum legal dose is 20 millisieverts, which statistically corresponds to a risk that one in every 1000 people who receive 20 millisieverts would get cancer.
- 9 millisieverts (0.009 Sv) total dose was received by the astronauts on the longest Apollo mission (Apollo 17, 302 hours flight time).
- 20 millisieverts (0.02 Sv) is the annual dose limit in many countries and the average dose that will be received by air transportation personnel (pilots, flight attendants) over about ten years.
- 50 millisieverts (0.05 Sv) is the hourly peak value within the Van Allen Belt behind aluminum that is three millimeters thick.
- 400 millisieverts (0.4 Sv) is the maximum permissible dose of for a person's working life in many countries and also of NASA astronauts, i.e. 20 times the annual dose limit. This increases the statistical risk of cancer to 20 per 1000 people.
- 2000 millisieverts (2 Sv) was the dose from the atomic explosion over Hiroshima within a radius of 1500 yards. Approximately 20% of all people exposed to this dose would die afterwards (however, 90% of the people within such close proximity to the site of the explosion died from the direct explosion effects of the bomb).
- If a person received more than 10,000 millisieverts (10 Sv), then he or she would surely die. For example, this corresponds to staying within a 19-mile

---

[2] For fun, I've gotten used to asking my doctors and their staff about the dose of radiation I'm receiving when I have to get an X-ray. This question almost always causes some unrest and commotion. One radiology assistant even said that the radiation would "degrade" over time. I switched doctors after that.

radius of the heavily polluted regions around the Chernobyl nuclear power plant for a year.

These figures show that even the astronauts of the longest lunar mission received significantly less radiation than is permissible in most countries each year. The crew of Salyut 6 received a peak value of about 55 millisieverts on their 4700-hour flight, almost three times the maximum annual dose for most countries. Not a single Russian or American astronaut exhibited significant radiation damage. Therefore, astronauts are not fatally irradiated as they fly through the Van Allen Belts. Since the orbital mechanics require a fast flight through the belts (total time in the belts of about one hour) to reach the Moon in a few days, no one can speak of any kind of special danger that the astronauts were in from radiation.

11

Too Hot, Too Cold

The Moon does not have any atmosphere and the lunar day lasts two weeks in Earth days. Because of these two facts, solar radiation reaches the surface of the Moon unhindered, and during the two-week-long lunar day, the surface can heat up to well over 100 degrees Celsius. It is truly a desert under the most extreme conditions. In order to protect astronauts in such an environment, every space suit made is effectively an all-in-one, climate-controlled cabinet with a window to look through. It isn't only the people who need the special protection of environmental control and life support systems, but also any objects brought to the Moon must be able to withstand the heat. Sensitive devices require extra care. This is especially true for the handheld cameras that are often the subject of skeptical questions, as they used to use very sensitive chemical films made of celluloid, which react poorly to high temperatures. The argument: **The celluloid film used during the Moon landings would have melted at such high temperatures. Therefore, it would have been impossible to take pictures on the Moon**.

This objection is understandable. Celluloid is a thermoplastic material that actually melts at relatively low temperatures, around 64 Celsius. At over 100 degrees Celsius, the Kodak-Ektachrome slide film used on the Moon is certainly unusable as a film medium. So how could you even take pictures on the Moon when the film would surely melt? (Fig. 11.1).

As with addressing previous points of criticism, a careful physical definition of terminology being used is required. If you want to talk about "temperatures" and frame the discussion in the right context, you first have to understand exactly what temperature is. When I say, "I am hot", everyone understands what I mean, but nobody knows exactly how hot I feel and how

© Springer Nature Switzerland AG 2019
T. Eversberg, *The Moon Hoax?*, Science and Fiction,
https://doi.org/10.1007/978-3-030-05460-1_11

Fig. 11.1 Charles Duke is working on a rock with his Hasselblad camera on his chest. Photo: NASA/J. Young. Product No.: AS16-116-18649

others perceive the same temperature. Of course, everyone knows that temperatures can be perceived differently. This perception varies from person to person, but also depends on other conditions that are out of our control. For example, the feeling of being cold is perceived as a function of air humidity. On one hand, it can be unpleasantly "wet and cold" and on the other hand we can also feel a "dry cold", which is much easier to endure.

Meanwhile, physics requires objective criteria that can be used to evaluate observations and that are valid in any situation. In this context, "temperature" is actually just another term for energy, and the energy being described is the motion of molecules. The faster the molecules move, the more energy they have. Therefore, in our daily lives we are actually describing the speed of molecules and their ability to transfer energy when we describe the temperature of something. When you sit in front of a warm stove, one way you can feel its warmth is through the warm air around it. Heat transfer is dominated by the contact of molecules that release their kinetic energy through collisions. Water is particularly good at absorbing and transferring heat, which is why wet and cold weather can feel like it "sucks" heat away from your body worse than dry and cold weather would.[1] Another well-known type of energy transfer is radiation, which we receive a significant amount of from the sun. The energy source is light: photons carry certain amounts of energy depending on the color of the light. Travelling from the energy source, the photons eventually strike an object and transfer heat to it.

There isn't any weather on the Moon, and in fact, there is no atmosphere at all to speak of. It wouldn't make any sense if someone were to describe a "muggy" heat on the Moon. Since the Moon lacks an atmosphere, there is no gas present to act as a heat transferring medium, leaving only radiation remaining as the source of heat on the Moon. It follows that as soon as radiation is eliminated as a source of energy, there is no longer any energy transfer to the Moon and it can quickly become cold. Therefore, the shaded, dark side of the Moon is freezing cold and the temperature transition at the border from unlit to lit side of the Moon is abrupt. The lunar astronauts, however, did not place their cameras on the hot lunar soil (which can reach temperature over 100 degrees Celsius). Instead, we have to ask about how hot it is around one meter above the ground.

The cameras still warm up without an atmosphere, but this process occurs very slowly—rays from the Sun are the only thing warming them up. They absorb radiated heat until a thermal equilibrium is reached between the heat being absorbed and the heat that is also being emitted away from the camera. This means that at certain temperature, the cameras would emit as much energy as they absorb. The so-called equilibrium temperature gets lower as more of the solar radiation is reflected away. If the camera is temporarily in the shade, then new heat is no longer being absorbed, while heat continues to be emitted away, further lowering the equilibrium temperature. In the case of the Hasselblad cameras, which were largely coated with a silver-colored paint that could reflect sunlight fairly well, the equilibrium temperature of the cameras was around 30 degrees Celsius. You may note that this is significantly

---

[1] The high heat capacity of water, i.e. its storage capacity of energy, is also the reason why we heat our homes with heaters that are filled with water.

lower than the temperature of the sunlit lunar surface and the melting temperature of the film.[2]

The solution to the risk of the camera film melting is thus actually quite simple once you ask yourself what you are actually talking about and then draw the proper conclusions.

---

[2] In contrast to the cameras, which could be put in the shade to cool down in the meantime, such a recourse was impossible with the un-air-conditioned lower stage of the lunar module (the upper stage as the astronauts' dwelling was actively cooled). Since it couldn't be put in a shadow, it was wrapped in a highly reflective Mylar film.

12

The Lander's Exhaust Plume and Its Crater

As the Lunar Module approached the Moon, it slowed down its descent by using a landing engine that generated a thrust of 45,000 Newtons. This amount of thrust corresponds to a weight of 27 tons on the Moon. After blasting the dusty surface of the Moon with so much force from landing the Lunar Module, surely you would expect there to be a crater created by the engine's exhaust plume. However, such a crater is nowhere to be found. This is confirmed in pictures from different missions where the ground below the descent stage is shown, even though astronauts reported the surface material to be as fine as flour. This results in the corresponding argument: **Obviously, they forgot to include an exhaust crater in the studio, even though a retrorocket engine firing hot gases into the landing area with several tons of force would have certainly made such a crater** (Fig. 12.1).

From the chapter on waving flags, we already know that critics of the Moon landings assume the Moon doesn't have an atmosphere. It turns out that, under this widely accepted condition, it is entirely logical that the landing does not create an exhaust plume crater. But why?

Gases flowing in a medium like air are generally turbulent. Their molecules collide arbitrarily and react at random. This effect is readily apparent if you watch someone quickly riding by on a bicycle. Their hair blows haphazardly in the wind, rather than standing straight backwards with no variation (a more fun way to notice this effect would be with a ride in a convertible, but that's beside the point). The reason that hair standing straight backwards looks so strange in some commercials is because we're all so used to turbulent wind. We simply cannot imagine turbulence-free flow (i.e. laminar flows) all that well. Turbulent flow also explains the chaotic and swirling clouds that are

© Springer Nature Switzerland AG 2019
T. Eversberg, *The Moon Hoax?*, Science and Fiction,
https://doi.org/10.1007/978-3-030-05460-1_12

Fig. 12.1 Buzz Aldrin with the Early Apollo Scientific Experiments Package (EASEP). Photo: NASA/N. Armstrong. No.: AS11-40-5927

made by the exhaust from Earth-based rocket launches. An impressive exhaust plume is emitted with a thundering roar as the rocket lifts off, serving as an illustrative example of turbulent flow and the corresponding sounds waves it generates. Without turbulence, you would only hear a quiet hiss as the rocket disappears into the sky. In the vacuum of space, however, there is no medium to resist the escaping gases, and as a result, the exhaust plume doesn't create any turbulence. All the particles in the plume move along well-defined streamlines, completely straight and silently away from the nozzle.

Two main components of a rocket engine include the combustion chamber, where the fuel and oxidizer are ignited and accelerated to high speeds, and an exhaust nozzle, which further accelerates the particles in a certain direction. In the vacuum of space where there is no turbulence, the exhaust plume is much more widely expanded than it is in an atmosphere. This is simply because with fewer molecules outside the nozzle to exert a pressure on the exhaust plume, the expansion of the plume continues after the exhaust gas leaves the nozzle cone. The result is that the gas has more possible directions of flow than it would with atmospheric pressure outside the nozzle. A considerable part of the exhaust jet is expanded to the point where it is ejected sideways. In turn, this means that any small crater that could be developed by the main, downward component of thrust, is blown away by the sideways components of thrust. Furthermore, due to the lack of an atmosphere, dust particles on the Moon don't experience any aerodynamic resistance. If dust gets blown away on the Moon, the particles fall back to the ground much further away than any of our experience on Earth would tell us.

Another aspect to consider is the landing procedure of the Lunar Module. The maximum applied thrust of approximately 27 tons was only used when the module had to be decelerated for the descent to the Moon from lunar orbit. Even though the speed necessary for orbit around the Moon is almost five times lower than that of the speed necessary to orbit the Earth, it is still almost 6000 kilometers per hour (3700 miles per hour). Just prior to landing, the Lunar Module was slowed down to the point where it would have appeared to be suspended above the surface, and the thrust from the engines was reduced to around two and a half tons. What most people don't know is that the descent motor was actually switched off at a height of 1.7 meters (5.6 feet) above the ground! This sounds quite adventurous, but in light of the reduced gravity, it isn't a problem at all. The turbulence-free exhaust plume and the subsequent freefall of the Lunar Module after engine cutoff are both shown in the corresponding film sequence from the Apollo 16 Moon landing. Moreover, the module did not always land vertically on the surface but floated down with a considerable lateral speed. When the landing module from *Apollo 11* threatened to land in a boulder field, Neil Armstrong took over manual control and flew over the boulders in an "extended" landing. Due to the unplanned nature of Neil Armstrong's correction, this final landing process was extremely stressful for everyone watching from Earth, leading Mission Control in Houston to throw out any other instructions they had for the landing process. In any case, this also led to the module landing with a decent amount of horizontal speed. One easy-to-spot effect of this is where the lunar soil is pushed up in front of the lander's feet (Fig. 12.2). As a result of the extended landing, the gas jet from the engine exhaust was spread out over the ground and only had a very brief effect on any individual plot of land.

The combination of no aerodynamic resistance on the dust particles being swept away, the pre-emptive engine thrust decrease and engine shutdown,

Fig. 12.2 Enlargement of the photo AS11-40-5927 with the lunar soil in front of the Lunar Module's foot

and the lateral thrust component leads one to the conclusion that an exhaust crater should simply not be expected. Such an exhaust-generated crater can only be created under the conditions present on Earth.[1] Therefore, a paradox is presented: Either there is an atmosphere on the Moon and flags flapping in the wind are completely normal, or the Moon has no atmosphere and the lack of an exhaust-plume crater is normal. Both arguments against the occurrence of the Moon landings contradict one another!

QR: The landing of *Apollo 16*. tinyurl.com/y7rz82aj

---

[1] There is widespread criticism that the landing feet of the Lunar Module are not completely covered by lunar dust. Allegedly, the dust would have been "stirred up" upon landing and then would have settled onto the lander's feet. This can also be explained by my considerations of the physics owing from the lack of air resistance and turbulence, and the pre-landing engine cutoff.

TOUCHDOWN ON LUNAR SURFACE

Fig. 12.3 Artistic representation of the Lunar Module including landing craters, made before the flights to the Moon. Photo: NASA. Product No.: S66-10992

There is one more very subtle argument that proves that the Moon landings actually took place. It can be found in the graphics that were published in advance of the landings. All of these pictures depict a crater caused by the lander's exhaust (Fig. 12.3). The photos of the Moon and the related drawings therefore become an argument for the Moon landings, and not against them. If they really wanted to fake a Moon landing, then why didn't they draw a scene with the landing crater they expected? Once again, the forgetfulness and ignorance of the technicians would have to be used to support such an argument against the Moon landings, as it was used in the argument of stars missing from the sky. Occam would have been delighted to see such contradictions and convoluted assumptions.

13

Anything Else?

The Spacesuits Are Too Stiff

Space is a vacuum. To survive there, humans need a pressurized vessel which maintains an Earth-like atmospheric pressure of 1 bar—a space capsule. And those who want to leave the capsule require a space suit equipped with a life support system and a pressurized atmosphere. At the same time, the space suit needs to be mobile and dexterous enough for an astronaut to be able to work while wearing it.

Space suits are actually quite bulky and stiff. An astronaut's freedom of movement is severely hindered, but of course they wouldn't want to go without it. The tools they use on the Moon and in Earth orbit are custom-designed so that they can be used with the necessarily thick space suit gloves. After all, the suit doesn't just provide environmental control and tolerable conditions for working in, but also protection from extreme temperatures and micrometeorites, which could strike you at up to 50 times the speed of a bullet. The climate control of the suit also cools the human body, which can start to sweat significantly during the strenuous work in a weightless environment. Several layers of plastic ensure that the suit doesn't overinflate or burst due to the significant pressure difference between the suit and the vacuum outside. But what about the gloves? They don't have any layers to prevent over-inflation. The argument is therefore: **Given the pressure difference of 1 bar, the gloves of the suit would correspondingly inflate to the rigidity of a car tire, and no one would be able to move their hands in them**.

This claim can even be substantiated in practice. Ralph René, a relatively well-known American critic of the Moon landings, built himself an air-tight

© Springer Nature Switzerland AG 2019
T. Eversberg, *The Moon Hoax?*, Science and Fiction,
https://doi.org/10.1007/978-3-030-05460-1_13

Fig. 13.1 Eugene Cernan trying on his space suit at ILC Industries. Photo: NASA. Product No.: AP17-72-H253

glove box with rubber gloves sticking into it. After vacuuming the air out of the box, it was very difficult for him to move the glove with his hand in it. Naturally, the developers of the space suit at ILC Industries in Delaware were well aware of this issue when they started designing the space suits for the *Apollo* program in 1962 (Fig. 13.1). They were able to solve it because the assumptions required for the above argument and for the test performed by René are wrong. In reality, suits and gloves are not just simple rubber balloons, but highly complex pieces of equipment. The gloves, as well as the space suits, had to be both flexible and capable of withstanding micrometeorites travelling faster than the speed of a bullet. This was achieved by making a pressure bladder that was molded from an impression of the astronauts' hands. An inner glove layered on top of the bladder was made of nylon dipped in neoprene. This layer was then reinforced to achieve a specified material stiffness. To provide flexibility, it was crucial that the internal pressure of the gloves was reduced to around 30% of normal atmospheric pressure. This allowed the surface tension of the material to correspondingly drop to the point where there was adequate flexibility for dexterous manipulation. Lastly, the outer glove cover was made of extremely expensive chrome steel woven into a fabric called *Chromel-R*.[1]

Some people would now claim that lowering the internal pressure of the glove so much would cause the blood in an astronaut's hand to boil, but that claim is false. Here, we can refer to Earth's high-altitude mountaineers. The pressure at the highest peaks of the Himalayas is around 30% of sea-level atmospheric pressure. While sometimes high-altitude climbers die from the consequences of breathing such thin air, it is never due to their blood boiling.

The Blue Windows

The pictures of the cockpit windows are referenced over and over again. In some pictures, they don't appear to be black, as one would expect when looking through them or at their reflection into space, but rather, they appear blue. Arguably, this cannot be true if the capsule was actually in space. **If you look through the window into space, the background should be black instead of blue. Therefore, the background must be the blue sky of the Earth, proving that the capsule never left Earth orbit** (Fig. 13.2).

The explanation for the blue windows can be found at the beginning of Chapter 4, where I explained how the blue component of white sunlight is scattered more strongly in the atmosphere than all other wavelengths (colors) of visible light. That's why our sky is blue. Light being scattered in the layers of the *Apollo* capsule's window behaves no differently. Just like the Earth's atmosphere, you can't see a black sky or a black reflection when sunlight hits a refracting medium (such as air or glass) and gets scattered by that medium (Fig. 13.3).

Sharp Footprints Cannot Be Made Without Water

Almost everyone knows of the famous footprint left on the Moon. The lateral grooves of the sole and its shape are sharply pressed into the dust of the Moon. However, the astronauts themselves have confirmed that lunar dust is as fine as flour. Due to the lack of water on the Moon, the question becomes: how

---

[1] Anybody interested in details about the development of the space suits, should check out *Part 5: Suits* of the movie series *Moon Machines* listed at the end of Chapter 15.

Fig. 13.2 Window of the command module of Apollo 7. Photo: NASA. Product No.: AS07-03-1557

could such well-defined footprints be made in completely dry soil? In other words: **The lunar soil is made of dust as fine as flour, and sharp footprints can only be made if the dust is wet. Since there isn't any water on the Moon, the footprints must not have been formed on the Moon** (Fig. 13.4).

I suspect that many children agree with me that the Moon is of course not made of flour, but it is more likely made of English cheddar cheese if Wallace and Gromit are to be believed. But if we don't trust these two friends, then we should start to consider the structure of the dust. Initially, lunar dust consisted of silicates, which were utterly pulverized by the impact of countless micro-meteorites. They are therefore not rounded like pebbles of sand, but they are erratically shaped micro debris with an extremely rough surface incapable of sliding past one another. There is nothing like it on Earth and the

Fig. 13.3 Astronaut Buzz Aldrin as he leaves the Lunar Module. Photo: NASA/N. Armstrong. Product No.: AS11-40-5863

analogy to flour cannot stand. Under pressure, the dust particles get stuck together and clump into stable shapes, even without the cohesive force of water to keep them together. Apart from the natural tendency of silicates to form molecular chains, the fractures on the surface of the dust particles do not oxidize, due to lack of oxygen, and can therefore become smooth due to weathering. The shapes of the particles are maintained until energy from a meteorite impact or the pressure from an astronaut's foot causes them to change again. If, in addition, the Moon's dust is statically charged, such as dust on a computer screen, by the permanent electron bombardment from the Sun, water is even more unnecessary to create and retain sharp footprints.

Fig. 13.4 Footprint on the moon. Photo: NASA/N. Armstrong. Product No.: AS11-40-5877

The Computer Technology

In this day and age, computers influence every aspect of our lives. For example, I'm writing this book on my laptop and I frequently head to the internet for research. The clock frequency of my computer is two gigahertz, and I don't need to worry about having enough storage space anymore. In the 1960s, computers were still in their infancy, while today, every car has more computing power and capacity than the Moon vehicles. **With their primitive computers, it should have been impossible to perform the complicated maneuvers that were necessary in space** (Fig. 13.5).

Most people are familiar with terms such as "operating system," "application software," "RAM," and "hard drive." But many younger people among

Fig. 13.5 Mission Control Center in Houston during the flight of *Apollo 14*. Photo: NASA. Product No.: AP14-S71-17122

us have never heard of the machine language "Assembler" or of toroidal core memories. Even with the older software and hardware, only 72 kilobytes of RAM (a decent modern computer has about ten million times more RAM) and a clock frequency of around 100 kilohertz (10,000 times slower than today's computers), one can still accomplish quite a bit. This is especially true if you are clever enough to separate the computing processes such that only the explicitly necessary procedures are performed by on-board software and everything else is performed by larger computers back on the Earth. In addition, routines were programmed to give priority to important calculations (e.g. the landing attempt) over less important ones (e.g. updating the air conditioning display). As every modern computer user knows, the result is that less computing capacity is needed. If you further consider that navigation was continuously supported by sextant measurements of stars, the Moon flights with the computers of that time are no longer so mystical (I refer here to the third part of *Navigation* of the film series *Moon Machines* listed at the end of Chapter 15). So, it is simply too short-sighted to compare today's computers and their computing capabilities and strategies with those of the Apollo computers. Just think of how much work and hard drive space a modern operating system needs, and how much is needed for compiling a programming language. If all the "dead weight" is shed, a computer from the "computing stone age" can also perform corrective maneuvers on its way to the Moon.[2]

[2] Robert Crippen, for example, knew this when he and his colleague John Young (the same Young, who made the jump on the Moon that I described earlier) made the first flight with a Space Shuttle. Crippen apparently did not entirely trust the computers on board the Shuttle. To be on the safe side, he took the then advanced HP41C pocket calculator on the journey. In case of a complete computer failure at on-board, he would have used it to calculate the ignition of the retrorockets for a safe return to Earth.

The Rover Has Problems

Wernher von Braun didn't just dream of a landing on the Moon, but also of permanent colonies. One of his ideas was to significantly extend the astronauts' radius of exploration on the Moon's surface after the first successful lunar landing. Of course, his idea was taken seriously because he was an essential figure in the lunar landing program. But it was impossible for the lander to visit multiple locations because, as with all spacecraft, the mass of the required fuel reserves for such maneuvers would be prohibitively heavy. And the extremely tight fuel reserves could not be significantly reduced (Neil Armstrong had less than 30 seconds of fuel left after his landing). Each additional ten pounds of weight (or 4.5 kilograms) reduced the available descent time by one second. Thus, for the last three Moon missions of *Apollo 15, 16,* and *17,* a so-called Lunar Roving Vehicle (LRV) was developed with the requirements that it shouldn't weigh more than 500 pounds (around 216 kilograms) on Earth and should be able to fit within the landing stage of the Lunar Module. The additional mass and corresponding shortened landing time could be compensated for by the accumulated landing experience, but the size of the rover was a serious problem. **The rover had a length of about 3.10 meters (10.2 feet), but the available volume in the descent stage could only accommodate half that length; therefore, the rover couldn't fit on the lander** (Fig. 13.6).

In reality, the problem was even more dramatic. As is well known, the landing legs of the descent stage had to be folded in for the launch from Earth so that the lander could even fit into the Saturn rocket under the Service Module of the *Apollo* spacecraft. It was hopeless to accommodate a vehicle more than three meters long and more than one meter high that also had wide, protruding wheels.

Two things have to be pointed out here. First, everyone involved had always understood that clever solutions would be required to get humans onto the Moon, and second, that solutions could often be found in the problems themselves. This would end up being a key insight for the engineers at Boeing and car manufacturer General Motors as they worked on solving the problem. From the outset, both their starting point and solution were outside the weight specifications required by NASA, as well as the available volume for the rover—and that, too, was dramatically small (Fig. 13.7).

The descent stage consisted of a double cross-shaped structure with a central square, which carried the engine. Square frames were also attached to its sides to accommodate various fuel tanks, so that the surrounding outer walls

Fig. 13.6 The lunar rover with a length of 310 cm (10.17 ft) in front of the Lunar Module. Photo: NASA/H. Schmitt. Product No.: AS17-141-21512

took the form of an octagon. Between the outer squares were triangular structures (so-called quadrants), which could also hold various other support systems and supplies. By reorganizing and moving some electrical components in the first quadrant, the engineers saw that they could use the resulting free space for the lunar rover. This meant that the engineers had a triangular column of space with a capacity of around one cubic meter (over 35 cubic feet) at their disposal. Certainly, a clever solution was needed here, because how else could an entire vehicle fit within such a small amount of space? The solution was found on the Lunar Module itself. If the legs of this spacecraft could be folded together within the Saturn for transport, then this principal could also be applied to the lunar rover, even if the latter problem is more complex.

Fig. 13.7 The folded rover during installation in the first quadrant of the landing stage. Photo: NASA. Product No.: AP16-KSC-71P-543

But engineers are passionate inventors not to be underestimated—their solution was as simple as it was amazing. The car was shrewdly folded so that it took the form of a triangular column and, together with an appropriate attachment mechanism, fitted straight into the bay of the first quadrant next to the exit ladder (Fig. 13.8).[3]

But the rover still hasn't quite pulled itself out of this affair yet. Its tire tracks are examined critically in several pictures, and if you look closely,

---

[3] The development of the rover and its components (folding mechanism, wheels, steering, cooling, etc.) as well as the ingenuity of the participants is impressively described in the sixth part *The Lunar Rover* of the film series *Moon Machines*.

Fig. 13.8 The Lunar Module of *Apollo 16* during descent to the Moon, taken from the command module. The first quadrant of the descent step is located to the right of the exit ladder. There, you can see the folded underbody of the Rover. Photo: NASA/K. Mattingly. Product No.: AS16-118-18894

something really does appear to be wrong. **The tracks of the rover suggest extremely sharp turns, which are impossible to make with a normal car. Furthermore, in many cases the second track of one of the rear tires is missing** (Fig. 13.9).

These comments are correct, but this is another case of a false assumption being made. The lunar rover is not a "normal car," but a special vehicle made for a very special purpose. If you look over the specifications of the rover (NASA has published them on the internet) and look at all the pictures of the vehicle, you will come across a particular detail that answers the questions from critics quite handily. In contrast to normal cars, each wheel of the rover could turn independently (Fig. 13.10).

Fig. 13.9 Charles Duke on the *Apollo 16* rover, the tire tracks show an extreme corner-ing and the usual second track from the trailing rear tires is missing. Photo: NASA/J. Young. Product No.: AS16-107-17446

During the planning phase of the *Apollo 15* to *17* lunar landings, it was clear that the areas to be investigated would lead to spectacular landscapes, in contrast to previous missions. *Apollo 15* landed in the vicinity of Hadley-Rille, *Apollo 16* flew to the Descartes Highlands, and the goal of *Apollo 17* was the Taurus-Littrow Mountains. In order to safely operate a vehicle in these areas, the rover had to be capable of flexible maneuverability on the ground. The concept of independently controllable wheels, which could be switched on whenever required, accomplished exactly that.

The rover had a minimal turning radius and the rear wheels ran in the tracks of the front wheels in this independently-controlled-wheel mode. This explains not only the extreme track radii in some pictures, but also the missing

Fig. 13.10 The parked rover. The picture shows that both the front and rear wheels can be controlled. Photo: NASA/E. Cernan. Product No.: AS17-143-21933

tracks of trailing rear wheels, which would be expected from normal vehicles without independently steerable rear wheels.

Ghosts in the Lens

In Chapter 4, we investigated why pictures from the Moon are void of stars and why the same is true for pictures from Earth orbit. We used the principle of "Occam's Razor" and, with the help of a plausibility analysis, revealed that it was both illogical and unconvincing to say that someone simply forgot to install small lights on a studio ceiling. In Chapter 6, we could see that additional studio lamps are not only unnecessary for non-parallel shadows, but they also are not a viable explanation due to the lack of secondary shadows (keyword: football game with floodlights).

Skeptics of the Moon landings, however, point further to apparently compelling proof of the studio recordings. The *Apollo 11* picture AS11-40-5872 allegedly shows two studio lamps shining in the upper left corner, with more beams of light from other lamps outside the frame. **Supposedly, this photo with studio lamps and their rays of light shining in the picture, was unintentionally published**.

And indeed, it would appear that the setting here is illuminated by several artificial light sources. However, there are two problems: First, we see that there is only a shadow cast from a single light source, which would be highly unusual considering the entire landscape is illuminated. Without any shadows from other spotlights, which would undoubtedly be necessary for lighting up the horizon (or, for all I care, the entire studio), the picture actually provides evidence against the argument that there are artificial lamps present. Then the question arises as to whether we are actually looking at artificial headlights here. To assess this, the high-resolution image can be enlarged to reveal the details.

The alleged lamps have a very unusual appearance at this magnification. Light not only seems to "flow out" from them, but they also have colored edges from red to yellow to blue, just like a color spectrum. The problem: pictures of headlights don't have colored edges like that. And because the astronaut, the lander, and all other objects are sharply in focus, we can exclude the excuse of a blurred image of the lamps. Still further, we can see rays of light running in different directions, which could indicate that they were generated by different light sources. But these rays are a shade of blue, despite the fact that the scene is by no means shrouded in blue light. So, what is going on here? (Fig. 13.11).

For reasons grounded in physics, optical lenses inevitably produce different errors within images. Because of this, simple glass lenses cannot capture an image where the entire field of the image is sharply defined. They produce a so-called "field curvature" in the image. The result is an image blurred at the edges or in the center. In addition, lenses have a focal length that depends on the wavelength of light they intend to capture and are therefore not able to reproduce all colors of an image sharply. When this occurs, colored edges appear on the objects that are shown in the image. To minimize these errors in photographs, camera lenses are assembled from several individual lenses made of different types of glass. This produces a much higher image quality of the subject being photographed. The physical process of capturing an image

Fig. 13.11 Buzz Aldrin sets up a sail to catch particles of the solar wind. Photo: NASA/N. Armstrong. Product No.: AS11-40-5872

through lenses is the called refraction of light in a medium. However, since there is no perfect lens, all camera lenses also produce unwanted stray light. The physical process of scattering in lenses is called reflection of light in a medium. Refraction, i.e. the direct image of the object, dominates the reflection (or scattering), because the selected lenses are designed to accomplish just that and (hopefully) they aren't scratched.[4] One thing to keep in mind is that the reflection or scattering effect increases with the intensity of the incident light. Even unwanted visible refractions can occur. Such effects often occur

---

[4] People who wear glasses with older, scratched lenses or glasses without an anti-reflective coating will be familiar with these light reflections.

when taking pictures outdoors into the Sun and while using wide-angle cameras with many lens elements, such as those used on the Moon. This effect can also occur when the Sun is not even in the field of view, as its light can scatter or break into the image at the surface of the lens. The result is that rays of light, rings, or light circles can expand over the image or reduce the contrast in the photograph. Stripes and secondary images like that are exactly what we are looking at in these photographs from the Moon (Figs. 13.12 and 13.13).

By also considering Occam's principle, we can conclude that the lights are not spotlights, and the diagonal blue stripes are not direct rays of light. Instead, the Sun is the cause of these phenomena. Hobby photographers know exactly what happened in these photographs. They will point out that they commonly have to deal with effects like these. There is even a term for it: "lens flare" in the camera lens. And given that blue light is more diffused in a medium than red light, physics also provides an explanation as to why the

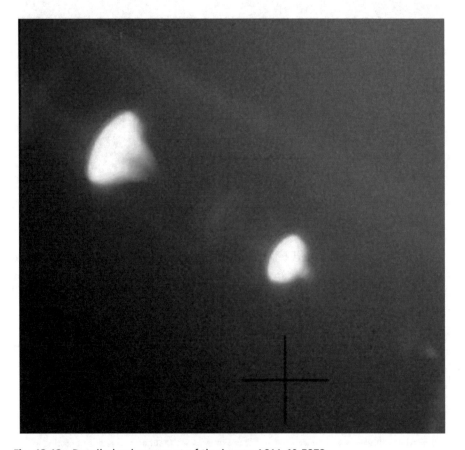

Fig. 13.12 Detailed enlargement of the image AS11-40-5872

Fig. 13.13   Detailed enlargement of the image AS11-40-5872

diagonal rays are blue. Once again, we have found that there is a plausible explanation for the supposed lamps in the picture through examples on Earth and with a bit of knowledge from photography and optics (Fig. 13.14).

Ghosts on the Screen

Those who, like me, also watched the first Moon landing on live television in 1969 may remember the strangely blurred and ghostly images broadcast from the Moon. The astronauts moved in front of the landing craft, the flags, and various instruments, but the objects were not immediately covered up by the astronauts stepping in front of them, but rather there appeared to be an afterglow effect of the objects remaining in the image. In light of the overwhelming amount of technology developed for the lunar landings, you have to wonder why such bad videos were transmitted to us, even though better image quality was clearly possible as evidenced by the pictures sent from the command capsule. Hence: **The live video was so bad because NASA wanted**

Fig. 13.14 Wide-angle backlit shooting with lens reflections. Photo: K. Vollmann

to cover up the fact that the Moon landings were recorded in the studio (Fig. 13.15).

Everyone wanted to watch live for the moment when Neil Armstrong descended the ladder form the lander and took his first steps on the surface of the Moon. This required an automatic camera mounted on the side of the descent stage, equipped with a very powerful wide-angle lens. At first, only a small antenna was available on the upper stage of the lander to send videos back to Earth, which transmitted in the S-band between two and four gigahertz (the command capsule had four of these). However, the transmission rate of this single antenna was relatively low because Armstrong had yet to set up the parabolic antenna that was designed for higher transmission rates. But if we had waited for him to do that, then people back on Earth wouldn't have been able to watch the first steps live. Transmitting the video with low data transmission rates and, thus, poor quality, was the only available option.

In order to be able to send the data back to Earth as video at all, the image resolution had to be greatly reduced. This left the video data in a state unsuitable for direct transmission to the television networks on Earth. The trick they used for sending the live video all over the world was to simply film the direct feed monitor screen in the Mission Control Center in Houston with a normal TV camera. Unfortunately, this led to the side effect of the monitor screen

Fig. 13.15 Mapping of a TV transmission sequence from the Mission Control Center in Houston

being mirrored in the camera lens and reflected back onto the monitor, where the TV camera recorded the secondarily reflected image again, triggering the "ghost like" effect.

The Rocket Man Meets Walt Disney

After the development of the atom bomb and the first rockets, the public was receptive to even the wildest of technological fantasies. Their thoughts were occupied by planets, the stars, robots, and of course, various aliens from space, which presumably would have to be fought with lasers. Hollywood produced a flood of new science-fiction films during this period of time. Business journals and magazines embraced and reinforced the developing enthusiasm for space travel. *Colliers Magazine* took advantage of this situation by asking the manager of the Redstone nuclear missile, Wernher von Braun, to collaborate on some articles where he would present his vision of manned spaceflight. The articles, embellished with spectacular drawings, were a huge success. That success was not lost on Walt Disney, who hired von Braun as a technical consultant for three television films about space travel. This cooperation has motivated the skeptics to argue the following: **If the Chief Engineer of the Moon rockets was working together with the most famous animator in the world at one point, then it is likely that they were actually working together much earlier to plan the faked studio recordings of the Moon landing**.

Unlike the theories discussed so far, this assumption has no technical aspect to it, which doesn't allow us to disprove it with physics. I'm taking it up anyway because it shines a light on what life was like in the 1950s and reveals how Wernher von Braun managed to convince the USA to turn his idea of Americans landing on the Moon into a real, funded program.

In contrast to von Braun's full-time work with the military, the articles at *Colliers* were a chance for him to direct public attention towards manned space travel. The idea of flying to the Moon was hardly on anyone's mind prior to the publication of his essays. When the articles appeared, Disney was just advertising its new Disneyland theme park in California. The park had four major themes, one of which, *Tomorrowland*, was particularly difficult to implement. What should be shown there, and how should it be shown? Von Braun's articles on space travel and those from natural scientist Heinz Haber on space medicine led Disney's producer Ward Kimball to hiring both of them, plus von Braun's colleague, Ernst Stuhlinger, as technical consultants for *Tomorrowland* and his TV series, *Science Factual*.[5] All three also appeared in the films themselves. Von Braun, Disney, and Kimball all understood that public opinion was influenced by television, and the collaboration easily benefited all parties involved. The three resulting films are worth watching, as they are quite visionary considering the state-of-the-art at the time. For example, Stanley Kubrick was enamored with the idea of a space station in Earth orbit for the leap to the Moon, and such a station can be found in Kubrick's film *2001—A Space Odyssey*.[6] Disney explained the scientific aspects of the work done by the engineers in his films. It was a wonderful collaboration. Von Braun developed the technical ideas, and Disney's artists gave them life. Science-fiction became reality (Fig. 13.16).

To take this collaboration as an indication of a conspiracy is to ignore the growing interest in space travel in the 1950s coupled with Wernher von

---

[5] The series consisted of the three television films: *Man in Space, Man and the Moon*, and *Mars and Beyond*. They can all be found on YouTube.

[6] In this context, critics like to refer to William Karel's award-winning film *Opération Lune/Dark Side of the Moon* think they can also prove a conspiracy with it. They miss the satirical character of this fictional documentary, which is full of half-truths and suggestions. The point of the film is to encourage us not to just blindly accept all assertions.

Fig. 13.16 Walt Disney (left) and Wernher von Braun at the Marshall Space Flight Center. Photo: NASA

Braun's aim to put his own dreams into motion. In addition, the necessity of explaining complex technical subjects to the layperson (a need which still exists today) shouldn't be overlooked. What right does anyone have to spend taxpayers' money on space exploration if no one in the public can even begin to understand how any of it works. To keep from cornering themselves into financial isolation, every expert must be capable of explaining the importance of their work. Scientists and engineers frequently do this and continuously attempt to improve public relations and educational outreach. Even today, aerospace engineers and science-fiction authors still commonly meet for public discussions and scientific programming can be found on TV. So, what makes the collaboration between Walt Disney and Wernher von Braun so mysterious? Disney sold tickets to his amusement parks and von Braun sold the public on his ideas. Today's television operates no differently—and so do conspiracy theorists, even though they are criticizing von Braun for doing the same. It isn't reasonable to attribute this collaboration to a conspiracy 15 years later. I regularly give lectures to explain astronomy. That is my job as a tax-funded scientist. And that's the same reason for why famous astronomer Vesto Slipher from the Lowell Observatory in Arizona took part in Disney's third television film *Mars and Beyond*. So far no one has criticized me or Slipher as astronomical conspirators and claimed that the stars don't exist. Although much of the ideas for future space travel were immature at the time, Disney's television series had far more substance than many critics think.

Science Factual was a great success. An incredible 42 million viewers watched the part titled *Man in Space*. Interestingly, shortly after the first broadcast, President Eisenhower announced that the United States wanted to launch an unmanned research satellite into Earth orbit to study geophysics in 1957. Von Braun was criticized in scientific circles, however, for publishing space travel propaganda in magazines and on television instead of publishing scientific articles in journals. However, von Braun was tasked with not only convincing scientists, but also industrialists, politicians, and most importantly, the public, in order to realize his dream of manned spaceflight. He understood the power of media and he used it. Ernst Stuhlinger described von Braun's commitment to his ideas as follows: "He fought on all fronts and each front had its own language. That was his genius."

Where Are the Pictures?

Despite my explanations, you shouldn't assume that everything can be clarified with logic and expertise. After all, it was humans who flew to the Moon and not infallible machines. In the prologue, I reminded you that NASA has apparently lost their original recordings of the first Moon landing. I still think that this is scandalous. How can you misplace the invaluable original poof of such a monumental event? It would be like the US Government losing the original Declaration of Independence. NASA still suspects that the recordings are in the archives of the Goddard Space Flight Center but cannot give a satisfactory explanation as to why the archival was not adequately documented. The only excuse is that archiving tapes during the Apollo era was a lower priority while the rockets were still launching. However, there is another, more worrying possibility: that the tapes were dubbed over in the 1970s and the original recordings were destroyed forever.

I imagine that it was more important for the creators of the Declaration of Independence to disseminate the document than to keep an original. But I must admit, given the loss of the Moon recordings, that I understand the ensuing criticism from all of the doubters about the Moon landing, and I cannot counter this argument.

Everything Is a Lie

In my arguments, I follow the widely recognized, epistemological approaches of natural science. But there are also other approaches to consider. The philosophy of solipsism describes the fundamental possibility that one's own consciousness is singular, i.e. that only your consciousness exists and no one else's does. Such an assertion cannot be refuted in the sense of a scientific-analytical approach, just as with NASA's lost data tapes. The idea that our world doesn't really exist was impressively staged in the Hollywood film *The Matrix*, but this concept of virtual reality is much older, and I cannot go into detail with it because of its complexity. It was implemented in quite a well thought-out and intelligent manner by the famous Polish science-fiction author Stanislaw Lem. A story in his novel *The Star Diaries*, published in 1961, describes boxes on an inventor's shelf, each simulating artificial consciousness and the entirety of its perceptions without the artificial character in them realizing their containment. Lem also explored the thought of what happens to the boxed consciousness when it recognizes the truth of its existence—it goes crazy.

14

Proof II: Rocks, Photos, and Stars

Now that we have highlighted the main arguments, it should be clear that the evidence presented against the authenticity of the Moon landings can be countered by both our everyday experiences and scientific analysis. In some cases, the assumptions that form the basis of the argument are already wrong. One should never believe that providing direct evidence of an event from the past is entirely hopeless, even if it occurred many years ago and on another planetary body. Despite the issues with evidence that I discussed in Chapter 3, you can at least find clues and indications that can be used to support an inductive analysis.

Distance Measurements

The distance measurements recorded by the *Lunar Laser Ranging Experiment* are a great example of indirect evidence of the Moon landings. As a result of this project, the distance between the Earth and Moon was determined with millimeter precision by the McDonald Observatory, the Halekala Observatory, the Observatoire de Calern, and the Apache Point Observatory. The principle for this proof is as simple as it is convincing. Using a telescope, a laser flash (an extremely tightly focused beam of light) is transmitted to the Moon. There, it hits a mirror which reflects it back to Earth. The distance from the Earth to the Moon can be determined by measuring the round-trip time of the laser flash from Earth to the Moon and back, and by knowing the constant speed of light (Fig. 14.1). The experiment showed that the Moon moves 3.8 centimeters (1.5 inches) away from the Earth every year. As simple as the principle

© Springer Nature Switzerland AG 2019
T. Eversberg, *The Moon Hoax?*, Science and Fiction,
https://doi.org/10.1007/978-3-030-05460-1_14

Fig. 14.1 The Goddard Space Flight Center's *Laser Ranging Station* with a laser beam pointing at the Moon. The Moon was overexposed to image the laser. Photo: Tom Zagwodzki/Goddard Space Flight Center

may be, two fundamental, non-trivial issues must be addressed for a successful experiment.

First, a highly efficient reflector must be placed on the surface of the Moon. You could illuminate the Moon directly, but because of its reflectivity (albedo) of 11%, it would be very difficult to get accurate measurements with such a correspondingly low amount of reflected light. This is even more difficult due to the fact that the light would be scattered in many directions while it is being reflected at the surface, significantly attenuating the signal. As to the second point, the reflected light also needs to be received back on Earth, so it can be measured. This is a real challenge! The beam of light from the laser flash spreads out on its way to the Moon, despite efforts to focus it to a narrow beam width (the tight laser beam from Earth has a diameter of approximately six kilometers, or just under 4 miles, on the Moon). It is reflected back by a relatively small mirror, and the reflected beam widens again on the return trip to the Earth before finally being captured with a relatively small telescope. Therefore, only a fraction of the originally transmitted light returns to the Earth (1 photon out of every 100,000,000,000,000,000 photons that reach the Moon), and telescopes are required to observe such a small amount of reflected light.

Measurements like these were already carried out before the Moon landings in the early 1960s, but there were large measurement errors due to the amount of light reflected from the Moon being too low for an accurate measurement. Since then, not only were the *Apollo* reflector mirrors used for successful distance measurements, but also reflectors carried by the Soviet lunar vehicles *Lunochod 1* and *2*.[1]

Now, it is often claimed that the mirrors on the Moon have to be aligned exactly to the Earth in order for the reflector experiment to work well. However, this is impossible. In this case, I agree that it is indeed very difficult to implement this experiment. But NASA never intended to place normal mirrors on the Moon. Rather, they used cleverly bundled mirrors containing up to 300 angled reflectors in the form of triple prisms on an aluminum frame, built by the German company Heraeus (Fig. 14.2). These prisms have the beautiful property that they reflect light back in exactly the opposite direction that it strikes them from, even if they aren't perfectly aligned with the incoming beam of light. Therefore, knowing the approximate positioning of the mirrors was sufficient for reflecting enough light back to Earth.

The quartz glass blanks were also supplied by Heraeus (Fig. 14.3). Quartz glass has a high optical homogeneity, and an advantage of this material is its relative insensitivity to high-energy radiation. Normal glass, on the other hand, reacts with such radiation over time, decreasing its transparency. Extraordinarily expensive space missions tend to avoid naïve mistakes, such as the use of simple mirrors. This is ensured by a highly complex management system. But it is also clear that without these well-designed reflectors on the Moon, there would be no reflection signals that would allow an exact distance measurement. These reflectors were set up by either humans, robots, or aliens.

---

[1] The *Lunochod 1* and *2* missions, relatively unknown in the West but very successful, were the first rovers on the Moon and were remotely controlled from Earth. I have already heard of the claim that in the early 1970s the Soviet Union was unable to develop a remote-controlled system and that both *Lunochods* were controlled by a suicidal Soviet soldier. I am leaving the judgment of such claims up to the reader.

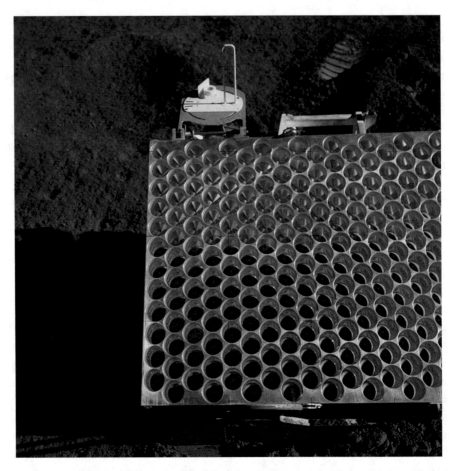

Fig. 14.2 Part of the *Lunar Laser Ranging Retro Reflector* from *Apollo 15*, just as it was left on the Moon. Photo: NASA/D. Scott. Product No.: AS15-85-11468

Either that, or every scientist who carried out the distance measurements and all the reviewers of the corresponding publications are liars.

Moon Rocks

A second, direct proof of the lunar landings is that of the rocks brought back from the Moon.[2] Together, all of the *Apollo* missions brought back nearly 400 kilograms (882 pounds) of rock to Earth, a considerable amount of which

---

[2] It is repeatedly claimed that the lunar rocks have been withheld from the public to this day. This is clearly not true, especially considering the permanent exhibits in museums around the world. In

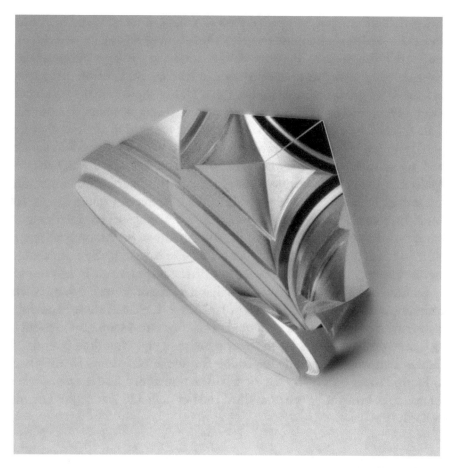

Fig. 14.3 Triple prism from Heraeus for the laser reflector on the Moon. Photo: Heraeus

have compositions that are unlike anything on Earth. The new names for these minerals include "Pyroxferroite," "Tranquilityite," (named after the *Apollo 11* landing area) and "Armalcolite" (named after Armstrong, Aldrin, and Collins). In addition, the natural isotopes Neptunium-237 and Uranium-236, neither of which exist on Earth, were found on the Moon. They exist there as a result of extremely prolonged bombardment of Uranium-238 with protons. Still further, very small impact craters were discovered on the rock samples, which can only be formed by the impact from

Germany, the Senckenberg Museum in Frankfurt/Main, the Deutsches Technik-Museum in Berlin, and the Haus der Geschichte in Bonn are some of the places where you can find such exhibits.

tiny particles from space. Impact craters like that cannot be found in samples on Earth, because such miniscule meteorites disintegrate as they pass through the Earth's atmosphere. The soil samples from the *Apollo* missions are still being tested by scientists to this very day. In January 2011, a team from Japan and the USA published new findings that clearly show that the rock could never have come from the Earth, as lunar landing opponents repeatedly claim.[3] Their investigations have shown that the proportion of "heavy" hydrogen (deuterium, which in contrast to normal hydrogen has an additional neutron in its atomic nucleus) in the water molecules is significantly higher in the *Apollo* rock than in any earthly water sample. It can't be from contamination, so the sample has to be lunar water. Such high deuterium values can also be found in comets, which provides not only a traceable source for the lunar water, but also further evidence for the authenticity of the Moon landings. In order to finish up the matter, skeptics would also have to find an explanation for the discovery of the isotope Helium-3. This isotope is made in and emitted by the Sun, but it is blocked from reaching the surface of the Earth by Earth's atmosphere. Rocks with inclusions of Helium-3, therefore, must come from space, such as those which were brought back from the Moon. The amount of rock brought back to the Earth totals more than eight hundred pounds, an amount that would be totally impossible for unmanned probes to bring back to Earth. Probes simply cannot do this. The unmanned Soviet *Luna* probes, for example, brought back a combined total of only 326 grams (0.7 pounds) to Earth (Fig. 14.4).

Radiowaves and Color TV

Many people are under the impression that the lunar missions could not be verified by institutions other than NASA. This, of course, is wrong. All of the radio waves that were emitted on the Moon were receivable by every ground station on Earth with the proper antenna. The director of the observatory in Bochum, Heinz Kaminski, became well known because his observatory confirmed reception of signals from *Sputnik* back in 1957. With their 20-meter satellite dish, the Bochum-based observatory was also able to receive signals from the Moon landings, and even archived them. Pictures broadcast on television, radio communications of astronauts with the ground station in Houston, and all other broadcasted flight data are still available there to this

---

[3] See: http://www.nature.com/ngeo/journal/vaop/ncurrent/abs/ngeo1050.html

Fig. 14.4   Moon rock, *Apollo 14*. Photo: NASA/A. Shepard. Product No.: AS14-68-9452

day. To receive the radio signals, the parabolic antenna had to be pointed very precisely at the Moon, and even the slightest positional deviations caused the signal to disappear (Fig. 14.5). This alone is proof that NASA would have at least had to place a relay station on the Moon in order to transmit a signal from there. To transmit such a signal from Earth or Earth orbit would have been impossible. This is because satellites in low orbits move too quickly over the ground, and the antenna receiving their signals would have had to have been adjusted very quickly to receive anything. With geostationary satellites, on the other hand, the antenna dish would not have needed to be moved at all. All other satellite orbits (e.g. geosynchronous orbits) are ruled out due to their complexity. More importantly, however, is the fact that the signals took about 1.3 seconds to travel from the astronauts to the ground station, corre-

Fig. 14.5 The Goldstone Deep Space Communications Complex (GDSCC) 70-meter parabolic antenna was used to communicate with the *Apollo* missions. Photo: UCSD

sponding to the approximate amount of time it takes radio waves to propagate from the Moon to the Earth. This rules out the possibility of signals being sent from Earth, relayed by the Moon, and then received back at the Earth—the transmission time would have been twice as long. Thus, the signals received on Earth could have only originated at the Moon, and nowhere else. The signal delay from Earth to the Moon and back can be observed in the video clip of *Apollo 15* astronaut Dave Scott falling over. The speaker in Houston says something regarding a "frame number" after 17 seconds, and you can hear the strongly muffled echo of the speaker from Scott's microphone at the 20 second mark.

Clearly it went to the Moon and then came back to Earth in exactly the expected amount of time. The fact that the signal was not simply stitched into the recording is evidenced by Scott's delayed reaction, which is synchronized with the radio echo that is heard back on Earth (he explicitly says the word "frame number"). The same thing happens after 28 seconds.[4] At Bochum and the other ground antennas that were focused on the Moon, only the signals from the Moon could be received, not those from ground control.

QR: The runtime of the radio signals. tinyurl.com/ybxhg5xa

Another story comes from the TV camera from *Apollo 12*. A camera developed by Westinghouse sent back live, color images to the Earth. As it was for *Apollo 11*, this camera was mounted on a descent stage instrument rack (Fig. 14.6). In order to transmit video of the astronauts' activities from the Moon, Alan Bean had to move it from the instrument rack to a tripod. In doing so, he accidentally pointed it directly at the sun, damaging the recording chip. That is why there are no color videos from *Apollo 12*. If everything had been recorded in a studio, a spare camera would undoubtedly have been available, allowing for the recording and dissemination of more convincing pictures of the Moon. Alan Bean's accident, however, is a striking example that the Moon landings happened. If recordings had been made in a studio, they would have almost certainly all been made in color.

QR: The failure of the color camera of *Apollo 12*. tinyurl.com/yd2fpwv5

These two examples show that the lunar missions, contrary to widely held beliefs, were not perfect. Three astronauts died during the early stages of the *Apollo* program, and on *Apollo 13*, the crew only survived thanks to good

---

[4] The way Dave Scott falls over, by the way, is another indication of the reduced gravity (see Chapter 8). On Earth he could have never caught his body so well above the ground. His reflexive leg movement during free fall also shows that slow motion is impossible.

Fig. 14.6 The color TV camera from *Apollo 12*. photo: NASA/C. Conrad. Product No.: AS12-46-6756

fortune. Why would you fake such accidents? These failures are indicative of how serious the Moon program was. Errors and deviations from expectations surprisingly help us verify the truth of the events, which applies just as much to the broken TV camera as it does to the missing exhaust crater below the descent stage (see Chapter 12).

Up to this point, the examples I have presented and analyzed thus far only shed a light on the primary doubts held about the Moon landings. I do not go into detail on addressing any further theories because they are simple to analyze. Among other things, this includes an Earth hanging in the sky (the photos were shot in orbit around the Moon). Thus, the subject of the Moon landing hoax should be sufficiently clarified by this point, except for two lingering issues: photos of the landers taken with telescopes and the infamous stars in the sky.

Probe Photos

In 2007, the Japanese mission *Selenological and Engineering Explorer* (*Selene*) was launched to the Moon. Among other things, the purpose of the mission was to create a map of the Moon and learn about its surface geology. The mission consisted of three satellites, which were able to perform three-dimensional surface mapping with the cameras they had onboard. One of the locations that was mapped was the landing site of *Apollo 15*, for which a 3-D topographic map was also built (Fig. 14.7). Because *Apollo 15* had landed in the topographically interesting region of the deep-cut Hadley-Rille canyon, the pictures taken in 1971 provide a very good comparison for the 3-D models developed by the *Selene* mission (the resolution of the cameras was too low for a direct picture of the Lunar Module, see Chapter 9). The result is astonishing! The pictures from the *Apollo* mission are a near

Fig. 14.7 The Hadley-Rille at *Apollo 15* landing site. Photo: NASA/D. Scott. Product No.: AS15-82-11122

perfect match with the results from the *Selene* probes (Fig. 14.8). The probes could resolve individual mountain ridges and summits, as well as the Hadley-Rille canyon, exactly as the astronauts saw them. You could now object, of course, that the *Selene* results are also fake. One should be careful with such an argument, however, because it fundamentally questions the work of modern science that we have all clearly benefitted from. It would also open the door to arbitrary speculation. I'll explain this in more detail in the next chapter. For now, suffice to say that the 3-D models from *Selene* not only support the authenticity of the *Apollo* images, but in return, the photographs prove the extraordinary performance of the technology used by the *Selene* mission.

As I already explained in Chapter 9, the *Apollo* equipment that remains on the Moon cannot be observed with Earth-based or orbital telescopes. An obvious and logical follow up question is then: why couldn't a Moon-orbiting satellite be built and launched to photograph the landing areas from the height of a few kilometers from the surface? The simple answer is that there is a lack of money available for such expensive space exploration projects. Furthermore, there are few people among those responsible for directing space

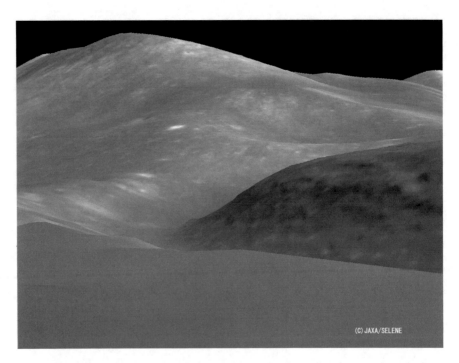

Fig. 14.8 3-D conversion of the *Selene* photos of the Hadley-Rille at the *Apollo 15* landing site. Photo: JAXA/Selene

projects who feel that a discussion about the authenticity of the Moon landings warrants such a follow up mission. In their opinions, more should come of the money spent on a project like that than just a few pictures for the skeptics. But in 2009, it happened that the Lunar Reconnaissance Orbiter (LRO) was launched to map the lunar surface. In addition to various instruments, the probe also had powerful telescopes and cameras on board. This was the first time that it was possible to photograph the *Apollo* landing areas in high resolution.

Naturally, space agencies operate their own public relations departments to ensure that the goals and benefits of expensive space exploration projects are clearly relayed to the public. Now, with the LRO, an appropriate opportunity arose to establish an answer to a question that has been widely discussed. The probe photographed the landing sites of every lunar mission.[5] Recordings not only showed the Lunar Module descent stages and their shadows, but also scientific instruments and even tracks of the astronauts (Fig. 14.9).

The Stars in the Sky

In Chapter 4, I explained why the stars had to be missing from the photos from the lunar missions. It was essentially because the astronauts were tasked with taking pictures of the scenery and landscapes on the Moon, not with making astronomical observations, or even taking such measurements. Owing to the following considerations, I have refrained from conducting an experiment which would thoroughly address this argument. In this context, there is already another proof of the Moon landings that could not be clearer. It takes astronomical observations and celestial mechanics into account.

Only a few experts are aware that astronomical measurements were taken during one of the flights and the results were extensively analyzed, evaluated, and later tested several times by the scientific community. Even as a trained astrophysicist, I did not know of these measurements for many years. An astronomical telescope for observing the stars in ultraviolet light (UV) was developed for the landing of *Apollo 16*[6] (Fig. 14.10). The telescope can be seen in Fig. 14.11, directly behind and to the left of Astronaut John Young. The

---

[5] Original recordings at http://www.nasa.gov/mission_pages/apollo/revisited/index.html

[6] The original publication can be found at http://www.opticsinfobase.org/abstract.cfm?URI=ao-12-10-2501

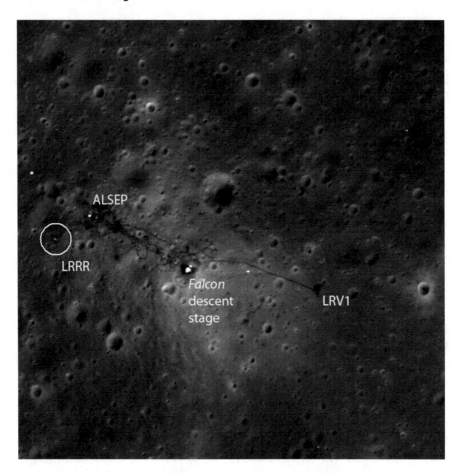

Fig. 14.9 *Apollo 15* landing site photographed by the *Lunar Reconnaissance Orbiter* (LRO). *The Lunar Laser Ranging Retro Reflector* (LRRR) is one of the instruments that could be identified. Photo: NASA/GSFC/Arizona State University

purpose of the measurements was to determine the composition of the interstellar medium, as well as to determine the mass of stars that radiate particularly brightly in ultraviolet light. Since our atmosphere and the hydrogen in the vicinity of Earth either block or outshine this light, such measurements are impossible from Earth,[7] which is why the telescope was designed to study those wavelengths in particular (observations in visible light can be performed on Earth, making it unnecessary to fly to the Moon for them). The telescope provided nearly 200 images of galaxies, nebulae, and stars that had never been

---

[7] For experts, this is the geocoronal emission in Lyman-alpha.

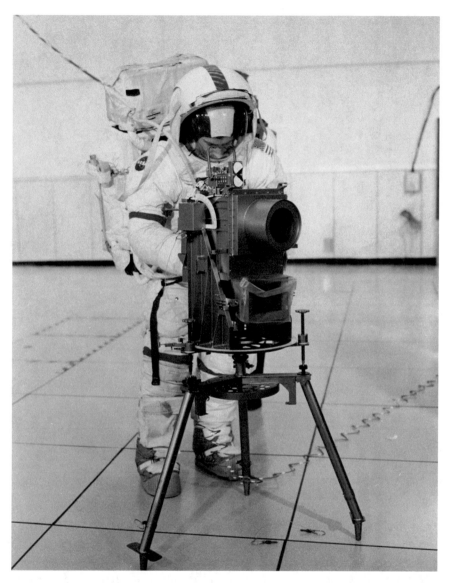

Fig. 14.10 Astronaut John Young training with the astronomical UV camera 1971. Photo: NASA. Product No.: AP16-S72-19739

seen before; a milestone for astrophysicists. Only two years later, the *Apollo* data was confirmed by the UV satellite *TD-1*. And after a veritable armada of telescopes was launched into Earth orbit, there is no remaining scientific doubt today that the results from the *Apollo* instruments were correct. The measurements were clearly made from space. Of course, one could logically

Fig. 14.11 Astronaut Charles Duke and the astronomical UV camera 1971. Photo: NASA. Product No.: AS16-114-18439

argue that a satellite orbiting the Earth took these pictures. But then they would need to explain why there is a strange sphere with odd stripes glowing in ultraviolet light in some of the pictures.

In reality, these very images are proof of the authenticity of the lunar landings. This can be shown by analysis with a star map. First, you have to determine which constellation is depicted (it is Capricorn), and then you can identify the stars in it. However, the luminosities of the stars are not quite in alignment with the known values from Earth observations of the same constellation. So, it is a manipulated data set after all? No, actually, the deviations in brightness are completely in line with what would be expected. Star maps refer to the light that can be perceived by the human eye, i.e. in the visible spectrum. Stars that appear dim in visible light can still appear much stronger in the ultraviolet light spectrum: it depends on temperature and some physical conditions of the star (dictated by the classified type of the star). The bright stars that make up the constellations are already well-researched and categorized. As it so happens, all the stars in Capricorn, which are now known to be very hot and emitting strongly in the ultraviolet light spectrum, appear very bright in the Apollo images. As previously mentioned, such results did not exist before and the unparalleled pioneering data could not possibly have been fabricated. These results still cannot be observed from Earth. So, at the very least, this data had to have been collected from space. But was it really recorded on the Moon?

The disc of light with the strange stripes is actually the Earth, which has been proven several times since then by later UV-imaging. The equatorial structures are excited atoms in the atmosphere, and they are not observable from a near Earth orbit, even with a UV telescope. It would have been an incredible occurrence if someone had imagined these structures and used them in a fake photograph, only to discover that they do exist. I'll admit that coincidences can happen, but even coincidences need to have a certain amount of probability to them (Fig. 14.12).

Fig. 14.12 Long-exposure UV image of the Earth in the constellation Capricorn. Photo: NASA. Product No.: AS16-123-19657

To shed some light on this "coincidence," however unlikely it is, we can still use the constellation of Capricorn and a star map to determine the place and time that the photograph was taken. Since the Earth moves relatively fast through the constellation (it passes through over the course of a few hours according to the orbital mechanics with the Moon as a reference point), we can determine almost the exact moment the photograph was taken. William Keel from the University of Alabama did just that. He simply used a modern, commercially available orbital mechanics simulation program; an experiment anyone can repeat. From NASA's web pages, you can find the data about when various work was carried out on the Moon, including the observation of the Earth in the Capricorn constellation with the UV telescope. With the time specified and the landing site of *Apollo 16* already known, you can see the position of the Earth at the same time with the simulation and compare it with the corresponding image from the telescope. As you will see, both in the original photo from *Apollo 16* and in the simulation, the Earth appears in the Capricorn constellation with an identical relation to the individual stars. This proves that the photograph was taken at the exact time and place as recorded by the astronauts in their logbooks. There are some stars in the original photo that glow much brighter or that have disappeared in comparison to the star map, but this can be explained by variations in brightness of visible light from the star map and the "invisible" ultraviolet light that was observed from the Moon.

QR: An analysis of the astronomical observations from the Moon. tinyurl.com/y9l6emzy

We also learned from this example that stars can, in fact, be seen in the sky from the Moon, as long as you adjust your photographic equipment accordingly. It is useful to note that there are even scientific investigations that are possible on the Moon that would be impossible on Earth because there is no atmosphere on the Moon to disturb the starlight being observed. Earth can also be used as a "position indicator" in case you get lost: by using a simple star map and observing where the Earth is relative to the constellations, you could determine the approximate time and your location on the Moon (this is exactly how high-seas navigators used to navigate by the Moon and stars from Earth). However, the data supplied from this UV telescope would not exist if the Moon had an atmosphere. Since there is no scientific doubt about the data obtained on the Moon, given the detail of the measurements and their later confirmation through tests done by others, there is clearly neither an atmosphere on the Moon, nor the possibility for flags to be flapping in the wind.

Taken together, the results from the distance measurements, the lunar rocks, the radio signals, the mishap with the *Apollo 12* color camera, the data from the *Selene* mission and the *Lunar Reconnaissance Orbiter*, and the astronomical measurements from *Apollo 16*, are compelling evidence proving that the Moon landings occurred. Moreover, we have seen that this evidence is capable of simultaneously invalidating several ambiguities at once. John Young's jump in front of the flag not only shows that no slow motion was used and that the jump took place in the gravitational field of the Moon, but you can also see the flag didn't move. So, at least during the filming there was no wind. And the video of Dave Scott's fall on the previous lunar mission not only depicts the correct timing of radio signal delays, but also like that of Young's jump, you can see that it does not occur in slow motion. This is the utility of Occam's Razor: a single explanation can answer several questions. And these questions being solved, in turn, answer other questions. The result is a harmoniously built argument without internal logical contradictions.

However, I will remind you once again of my remarks made in Chapter 3: If scientific methods are not accepted by the critics, then we can never find meaningful proof in the scientific sense. To put it another way, if the above evidence is also considered to be faked, such as the distance measurements or probe photos, then a discussion about the reality of the Moon landings is fundamentally impossible. All that would remain would be for every critic to visit our Moon so that they can see the truth for themselves.

In summary, I investigated some critical questions regarding the Moon landings and I have arrived at the end of my considerations. However, some fundamental questions arise from theories, analyses, and new contradictions, which I will take a closer look at in the next chapter.

15

What Can We Learn?

The Moon landing was certainly an extraordinary event that forever changed the way people saw humanity as a whole. Not only did all of the new technology surpass our wildest dreams, but we also gained a new awareness of our place on Earth in a global, shared community. One of the most profound experiences that the lunar astronauts reported did not come from viewing their destination, but rather taking a look back on the "fragile" Earth they had left behind (Fig. 15.1). In hindsight, many astronauts say that on our way to explore the Moon, we discovered the Earth instead. The Moon landing was a giant leap forward for humanity, but it often goes unrecognized just how rapidly the world has been evolving since. Our lives became incredibly more complex and driven by technology. Overall, people are amazed by all of the advancements, and therefore it is natural that doubts and criticisms may arise. I consider skepticism to be a virtuous quality that helps you to sharpen and develop your mind, leading to a higher quality of life. However, skepticism requires an investment of time and effort to be useful. The sheer consumption of claims does not make someone any more knowledgeable and has a devastating effect in a confusing world. An attendee at one of my lectures on the subject of the Moon landings once asked me to speed up the presentation because he thought my explanations seemed to extensive. But to do so would have jeopardized my ability to calmly and thoroughly address the doubts of the Moon landing skeptics, including his own, in addition to presenting my analytical methods.

At this point, I believe it is appropriate to discuss the origins of the Moon landing conspiracy theories. This is necessary because the authors of these theories that I have discussed are now considered experts since publishing

© Springer Nature Switzerland AG 2019
T. Eversberg, *The Moon Hoax?*, Science and Fiction,
https://doi.org/10.1007/978-3-030-05460-1_15

Fig. 15.1 Earth as it was seen by *Apollo 17* on its way to the Moon. Photo: NASA/E. Cernan. Product No.: AS17-148-22726

books on the subject. The theories, which I have exhaustively analyzed, have become a part of pop-culture, and thus influence how we and the media see the world. At one point I learned that the American author, William Kaysing, who has a Bachelor of Arts in English, published a book titled *We Never Went to the Moon: America's Thirty Billion Dollar Swindle* in 1976. In it, he claimed that the technology necessary for a Moon landing did not exist at the time, so the flights could have never occurred. He was the first to point out how stars were missing from the photographs from the Moon and that the lengths of shadows seemed strange, which led him to conclude that there must have been a conspiracy perpetrated by the US government. He claimed that all of the recordings and evidence were held in the restricted military area called

"Area 51,"[1] and that he himself had survived several assassination attempts by the CIA.

Kaysing, who died in 2005, was no fool. He had authored several non-fiction books, and as a former head of technical documentation to one of NASA's supplier companies, Rocketdyne, he was well informed about the American lunar program. His book on the Moon landing wasn't just a blip on the literary radar, but rather it had considerable public impact. Respected television stations debated his theories, and at the same time, made them known to a wider audience. Nevertheless, as my considerations in this book reveal, I retain a great skepticism against his allegations.

Unfortunately, Kaysing's theories were only critically examined by a few authors and editors, allowing his arguments to persevere in society worldwide. Meanwhile, a whole series of books, websites, and other articles were published about the Moon conspiracy. Most of them perpetuate Kaysing's theories without doing any research into whether or not they are true, in addition to adding their own unverified input. Such is the case with the claim that various astronauts, who were designated as secret-bearers, were killed by NASA. This particular claim was staged in quite an exciting way in the Hollywood-produced movie *Capricorn One* (I like this movie, really).[2] The constant repetition of claims does not increase their truthfulness, which only a comprehensive analysis could do. However, the repetitions alone have been enough to influence internet search results to list the conspiracy theory sites before sites that refute them, even if one is only searching using the term "Moon landing." Obviously, the topic has penetrated public life to such an extent that its significance is almost equal to that of lunar missions.

Please don't misunderstand me, I still believe these theories should be taken as legitimate challenges in an open and scientific discussion. I welcome them! If they are not only intended as entertainment, then critically examining them should provide us with a net gain in knowledge about the subject. If a theory is refuted, then the claimant should not simply drop it and blindly turn to the next argument, but rather critically evaluate the "evidence" that they presented and their approach to presenting it. This method of self-evaluation has been a common approach throughout the history of scientific development and especially helped contribute to the age of scientific enlightenment in the 18th century. It is far easier to make and spread claims than it is to analyze and

---

[1] "Area 51" in Nevada as part of the Nellis Air Force Base is itself the subject of various conspiracy theories.

[2] One may ask why Kaysing died a natural death at the age of 83 years only in 2005, almost 30 years after his "revelations".

evaluate their content. This becomes particularly important when competence and technical expertise are required for the thorough evaluation of a claim.

What comes from ignoring the methods, history, and successes of science can be seen in the "findings" of Ralph René, mentioned in Chapter 13. In addition to false claims about the dexterity of the astronauts' gloves, he questioned the validity of Newton's and Einstein's theories of gravity, Archimedes' principle of buoyancy, and Coulomb's law of electrostatics. He was also convinced that the solar system is not held together by gravity, but by electrostatic forces. He criticized the fact that his research was not published by any reputable research journal, thereby accusing established science of suppressing his ideas. The accusation is legitimate. However, I would like to point out that scientific investigations, such as the laser distance measurements and the geological findings of the lunar rocks, undergo strict peer review procedures, i.e. scientific cross-examination by external scientists. This proven method certainly has some weaknesses and should always be double-checked. However, it has contributed greatly to our technological progress since being implemented. Scientific knowledge is always related to the knowledge gained up to the present day, even if older knowledge is already discarded. The philosopher Bernhard von Chartres had this insight for the first time in the year 1120 when he stated "… *we are, as it were, dwarves sitting on the shoulders of giants in order to be able to see more distant than them—of course not thanks to our own sharp eyesight or body size, but because the size of the giants raises us up.*"[3] Not only did he express his admiration and respect for the old masters of antiquity, but he also recognized their significance in context of the progress of science through history. Einstein acknowledged this fact by regarding his own theory of relativity as merely a "modification" of Newton's theory of gravity. And Newton, in turn, saw himself as a "dwarf standing on the shoulders of giants.[4]" Ralph René on the other hand, who died in 2008, had always negated his dependence on science from the past. When asked about his education, he always referred to his knowledge as being self-taught and implicitly highlighted its lack of grounding in historical truths from the old masters.

I've only heard of three reactions from *Apollo* astronauts to the conspiracy theorists' allegations. In the 1990s, Jim Lovell (*Apollo 8* and *13*) said he considered Kaysing's theories to be "wacky," which was the basis of a slander lawsuit that Kaysing pursued against Lovell, although the case was dismissed

[3] John of Salisbury, *Metalogicon*, Publisher: John B. Hall, published 1991, page 116.
[4] Letter to Robert Hooke, February 5th, 1676.

in court. Bart Sibrel, and American filmmaker, went even further with his actions and undermined any critical discourse he could have had with his behavior. When he asked Neil Armstrong (*Apollo 11*) to swear on the Bible that he had been on the Moon, Armstrong declined quite rightly, remarking laconically that his Bible might be a fake (Sibrel gave the meager reply that the Bible was actually real). By far the most striking response came from Buzz Aldrin (*Apollo 11*) in 2002. Sibrel also asked him to swear on the Bible, and when he refused, Sibrel suddenly and unexpectedly insulted him, calling him a coward, a liar, and a thief! Sibrel was obviously not aware that he could have gotten himself caught up in a libel lawsuit. Aldrin, on the other hand, had written his doctoral dissertation on orbital rendezvous maneuvers, and I like to think that in this moment he was teaching a lesson about the Newtonian principle of "Action and Reaction." The 72-year-old punched Sibrel in the face with a strong right jab,[5] which caused quite a stir in the media, and the video of the event quickly became world-famous.

QR: Buzz Aldrin meets Bart Sibrel. tinyurl.com/yd6elc4l

NASA refrains from making any rebuttal to skepticism about the Moon landings. Their passiveness is a simple indication that they believe the burden of proof lies with anyone who believes the conspiracy theories. Instead of logically substantiating their own claims, the conspiracy theorists cite this behavior as reason to believe that NASA has something to hide. Further unsubstantiated claims include that NASA suddenly lost the blueprints to the *Saturn* Moon rocket after the *Apollo* program and that the Moon rocks were never real.

You can't learn anything without hard work. Ernst Stuhlinger, one of the key figures in the development of the American lunar program once said: "The way to *believing* is short and easy, the way to *knowing* is long and rocky." He was referring to the need for education and perseverance in order to understand anything in an increasingly complex world. It is human nature to be easily distracted, and it can be truly exhausting just to keep yourself from indulging in relaxing entertainment. I must confess, I am a fan of the TV

---

[5] All construction plans of the *Saturn V* rocket are stored on microfilms in the archives of the National Space Science Data Center.

series *Star Trek: Enterprise*, even with all of the nonsense in it, but it isn't too much of a distraction for me. However, when entertainment causes confusion and uncertainty, then it can have a negative effect on your quality of life. As early as 1784, Immanuel Kant once wrote[6]: "*Enlightenment is Man's departure from self-inflicted immaturity. Immaturity is the inability of one's mind to understand without the guidance of another. This immaturity is self-inflicted if the cause of it is not the lack of reason, but of resolution and courage, with which to help oneself without being led by someone else. Sapere aude! Have the courage to serve your own mind! This is the motto of Enlightenment.*"

Neil Postman accurately described this issue in his 1985 book about media in society, *Amusing Ourselves to Death: Public Discourse in the Age of Show Business*. He conjectured that we were making the move from a content-oriented to an entertainment-oriented society where reflection on the conveyed content is either discouraged or hindered. He warns against the passive, uncritical consumption of media, because a stable society depends on the ability of people to be discerning of information. He saw that it poses the risk of putting society on a path towards totalitarianism, as imagined by Aldous Huxley in his book *Brave New World*, in which people are not suppressed by the government, as in Orwell's *1984*, but by the consumption of entertainment. Postman's concerns are not without basis. Kant has to say on the subject: "*Laziness and cowardice are the reasons why the greater part of the population has long since resigned itself to being led by others (naturaliter mariorennes), remaining blissfully immature for their entire lives; and these are the same reasons that others find it so easy to control them. Ignorance is bliss.*"

Nevertheless, conspiracy theories (the Moon landing is only one example from many) are experiencing a boom right now. They are spread and shared over and over, becoming ever more enriched with untruths. I see analogous behavior to the passive consumption described by Postman all over the place. Just like advertising, if you keep repeating the same thing often enough, people will believe it. But the motives of conspiracy theorists are not clear. If it was for financial gain through the sale of such theories, then that wouldn't be a crime. However, the respective authors usually reject this outright and declare themselves to be the bearers of truth to an ignorant world. If this is the case, then it betrays a frightening lack of analytical education. What we can be certain of is that such claims propagate an interpretation of world history that creates more uncertainty than it does clarity. It makes no difference whether the claims are made about aliens using the pyramids of Giza as

---

[6] Berlinische Monatsschrift, December issue 1784. Pages 481–494.

landing signals, despite a clear and thoroughly researched Egyptian history, or about the unforgivable terrorist attacks in the USA on the 11th of September in 2001 being part of a government conspiracy.

The psychologists Michael Wood, Karen Dougles, and Robbie Sutton from the University of Kent have been investigating what makes people believe in conspiracy theories despite clear facts against them.[7] In a survey of psychology students, they found that some people can even regard mutually exclusive explanations as being simultaneously plausible. This presents a very strange survey result. The belief in the assertion that Princess Diana faked her own death correlated significantly with the suspicion that she was murdered! This brings us to the conclusion that the content of a conspiracy theory is unimportant. Rather, the general belief simply needs to fit into the worldview or with the ideas that the person already has. We call that "prejudice." The authors reference similarities to the studies conducted by philosopher and sociologist Theodor Adorno with people who hold anti-Semitic beliefs that are contradictory to one another and suggest that people who exhibit certain ideological traits are pre-disposed to believing in conspiracy theories.

The astrophysicist Harald Lesch had this to say about the subject: "*We know … that we cannot understand every aspect of normal life because things have become too complicated. Every one of us is ultimately dependent on trusting other people … Behind the perception that an event like the Moon landing is a conspiracy, there stands an associated worldview that can only be described as a deep-seated mistrust. People trust others to do things that they know they obviously would trust themselves to do. They would have done the same in this case.*"

This detail is an interesting aspect of what is probably the largest, most recent conspiracy theory—the terrorist attacks of September 11, 2001 in the USA. I don't want to go into too much detail with this theory, but with a little research and logic, I can show you how to dissect one argument from this theory. The argument: **Nothing remained of the airplane that allegedly crashed into the Pentagon. The attacks on the Twin Towers and the Pentagon were really perpetrated by government agents to destabilize society.**[8]

---

[7] Social Psychological & Personality Science, 2012, DOI: https://doi.org/10.1177/1948550611434786. The publication can be found at http://m.spp.sagepub.com/content/early/2012/01/18/1948550611434 786.full.pdf

[8] Such an "uncertainty" through a government conspiracy was not at all necessary considering what happened after the attacks, as terrorists had already achieved this through earlier attacks. Two wars were instigated, civil rights were ignored, and freedoms were restricted. Even to this day, every foreign visitor to the USA today is observed during entry into the country as if they were a criminal. These "measures" enacted by the US government were supported by the majority of US citizens. If all this was an act of the

So where did the plane go? Did it disappear? Even though the walls of the Pentagon were built with bricks, the outer walls were reinforced with steel beams, built around the bricks. There are more steel columns further inside the building. Meanwhile, the airplane would have crashed into the building with the kinetic energy of over 4000 Volkswagen Beetles travelling at maximum speed. One thing is clear: a Boeing 757 travelling that fast could break through the walls of the Pentagon without a problem. And because the airplane is made in part with soft aluminum, it was easily shredded by the steel beams (you can see exactly that happening in video recordings made of the crash at the World Trade Center). Once the plane was inside the building, the fuel exploded.

All of the aircraft had only taken off a short time before their respective impacts, so the fuel tanks were still filled with around 40 tons of jet fuel and each aircraft was relatively heavy, weighing approximately 120 tons. The speed of each plane at impact would have been close to 700 kilometers per hour (435 miles per hour). Taking these values and the energy available in kerosene, and using some high school physics, you can figure out that the jet fuel would have had 750 times more chemical energy than the kinetic energy of the fully refueled aircraft. This shouldn't be surprising, as there needs to be a lot of stored energy for the fuel to help keep the aircraft in the air for many hours. The energy content of the jet fuel corresponds to the energy of 350 tons of TNT. With that, the explosion that went off in the building did so with the energy of a small nuclear bomb. It isn't a wonder at all that an aircraft, which consists primarily of aluminum, would be completely destroyed in such a way that nothing of it remains. But in reality, not all of the larger components of the aircraft (engine turbines, parts of the landing gear, etc.) were completely destroyed. Some of them were later found, and you can locate pictures on the internet of landing gears and engines that remained in the wreckage on the streets after the attacks. Together, these facts explain how aircraft could penetrate into the buildings with their kinetic energy, and then cause such enormous damage inside from the chemical energy of the jet fuel. As you can see, this is just one brief analysis to refute the arguments of conspiracy theorists.

Therefore, the false allegations about the attacks in the USA are not interesting. These were despicable, criminal acts, but they should not have led to wars. If anything, they should have caused increased compliance with the "Montreal Convention" and unification of certain rules on international air

---

government, one may ask, why every ridiculous "sex affair" of a US president becomes public knowledge, but not such a gigantic fraud.

transportation therein. What I find more compelling to discuss is the question of why some people create and propagate conspiracy theories, and what they aim to achieve with their allegations.

There is one fundamental and all-encompassing assertion that I do not want to ignore. It goes: "**Yuri Gagarin's space flight was also faked and the USA, relying on the secrecy of the Soviet Union, was able to fake its own Moon landing eight years later.**"

If this were true, could any of the governments around the world be trusted? This theory might seem easy to brush aside because, in its enormity and its monstrosity, it seems impossible to refute. That is, in the sense of scientific proof for a historical event. Still, there are events from history that give us an idea as to whether we have cause for concern. It is possible that someone could ask whether the Cold War was real, or if it was just a harmless ruse put on by two super powers. The several thousand nuclear weapons, which could have wiped out mankind several times over, could have only been made as the result of a mutual agreement. As someone who lived through the Cold War and saw it for himself, I consider such an argument to be made carelessly. It belittles a situation in which several million soldiers were armed to the teeth and ready to attack the opposite side at a moment's notice in an emergency. And they would have done so using weapons with destructive power beyond imagination. For the young readers to get a better idea of what that was like, albeit a small one, I would refer them to more recent powder-keg that is the border between North and South Koreas.

History shows that spaceflight has commonly been used and regarded as propaganda with which to demonstrate the superiority of the respective political system while also to discretely showcasing the capabilities of related lethal weapon technology. The Soviet Union, like the USA, also developed a lunar rocket. The *N1* was around the same size as the American *Saturn V* and had a similar take-off mass but was capable of approximately 30% more thrust. Little was known about the *N1* in the West until the collapse of the Eastern Bloc. Only then did the general public learn of the failures of all four launches of the *N1* with unmanned capsules that occurred between 1969 and 1972. The Soviets were very much in the race for the "Moon Cup," and the American engineers could not simply sit back and coast to victory. Today, we know that the technical reasons for Russia's failures lie in the complex engine technology. The cause of this, however, was that two teams of Soviet engineers were separated and competing with one another due to their disagreements about exactly how to develop the engines, thereby not sharing knowledge or working together. It is an interesting question as to how the race for the first Moon landing would have turned out if the Soviets had used a single, coordinated engineering group to develop the engines for the *N1* rocket, rather than two severed groups shrouded in secrecy from one another. Therefore, the thought of a "non-disclosure agreement" between the USA and the Soviet Union during this era is completely absurd. Why would the East have tolerated a conspiracy perpetrated by the Americans when they themselves failed spectacularly in the race for technological and political prestige?

History is never made without context, and the race to the Moon can only be understood in the context of the political developments at the time. These include such critical events as the Cuban Missile Crisis, the construction of the Berlin Wall, or the USA's Vietnam War. You have to take the connection between these historic events into consideration if you are going to postulate a theory of conspiracy between the United States and the Soviet Union. For some people, even my explanations are now a part of the conspiracy. I have been "flipped" to be a part of the great, world-wide lie since I began studying physics. Naturally, I have been well-trained to feed my false, physics-based arguments to the readers to convince them of the lies. I no longer have any response to such accusations because they attack my personal integrity. This is the manifestation of a fundamental mistrust in the world and its inhabitants, and so civilization itself is negated. It is a suspicion that accuses people of fraud on the basis that people would do something like that because they are all villains.

I can live with that because I simply do not deal with people who attack my character. The behavior of such people usually would not be a problem if it weren't for the multipliers, i.e. the media and their representatives, whom adopt and disseminate unchecked assertions in favor of sales and ratings, largely shirking their civic responsibility as the "fourth branch of government" (fourth, after the executive, legislative, and judicial branches). This is exactly what occurs when the media voluntarily synchronizes around an idea like the "War on Terror" in the USA, a practice that is threatening to establish itself in Europe. This is precisely the threat to a democratic civil society that Neil Postman warned us about, and to which the astronaut Buzz Aldrin says, "*I think that people whom deliberately mislead young people, meaning the future decision makers, to their own benefit, should be held accountable for their actions.*"

Doubts about historical events will continue to be raised in the future. This is good in the sense maintaining the critical thinking that only serves to strengthen a discerning society that is open to discourse. However, this must include good research, logic, and common sense. The world is so inscrutable and complicated that people are always looking for ways to get out of the forest of uncertainty, just so they can get their bearings back. And so, books that

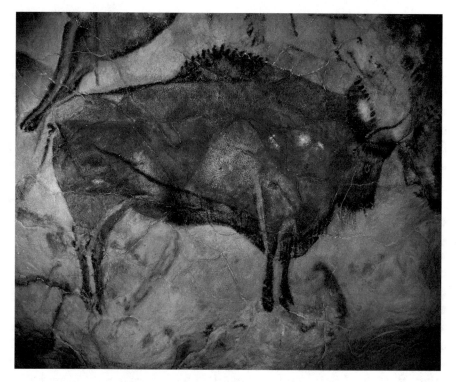

Fig. 15.2 Mural painting in the cave of Altamira. MNCN-CSIS, Spain

are published with nonsensical content that cater to this desire to "see behind the curtain" will always make money somehow. It does not matter whether it's about the Bermuda Triangle or the arrival of aliens in Peru. Regardless, we should always dare to contradict and disagree with each other. This doesn't have to be on all topics, but it is enough to delve a bit into any single topic and cultivate some skepticism.

Nevertheless, the success of the "Moon landing conspiracy" cannot be denied. So, what is it that makes people believe in claims that fall apart when viewed rationally? Why is it that people do not trust in scientific explanations and analytical studies, but rather believe in theories presented without proof? I have already discussed the differences between passive consumption and active questioning of media, but this cannot be the only explanation. I've been inspired to pursue an answer to this question by people who live their lives with deliberation and keep themselves from making passive decisions or allowances. I believe that the fertile soil in which the Moon landing conspiracy is planted is made possible by a modern world dominated by knowledge. Human history is full of Beliefs and Myths that gave us direction and purpose in the world (Fig. 15.2). If instead, everything in the world has a rational explanation, then the very existence of such myths is threatened. It seems to me that people resist logic in favor of spirituality. That isn't to say that I'm thinking about how the Sun is just a large fusion reactor while I'm watching a sunset—I, too, can find beauty and wonder in the natural world. These kinds of feelings are what distinguish us from the animals. In this respect, it would be a fatal error to disregard beliefs and modern myths, such as ghosts or aliens landing on Earth, as pure nonsense. We should accept that the desire for stories and understanding of the inexplicable goes back to the first cave drawings made by *Homo sapiens* in the Ice Age, and still resides within us to this day.

A major problem arises when new myths or bases for faith are presented as scientifically valid. In contrast to religions, which have never claimed that spirituality can be explained or quantified by science, representatives of some modern religions try to support their views with science. The inherent failure of technology and UFO conspiracies to scrutiny is inevitable. The natural sciences cannot "prove" any aspect of spirituality or faith, so their tools of reason and deduction cannot be used disprove them either. For example, I cannot say anything about my faith from a scientific point of view. The monk, William of Occam, was well aware of this, and had no problem maintaining his faith while at the same time rationally observing the world around him, just as priests from the Vatican continue to work in the field of modern astrophysics. In my opinion, it isn't even a problem if you don't believe in the Moon landings. However, it is best not to attempt to use analytical methods to validate such a belief, because analysis and deduction in this case will lead you to conclusive answer.

Even if proponents of the conspiracy theory allege that a majority of people believe in the Moon landing conspiracy,[9] it does not mean that the "majority" is right. The truth cannot be decided upon by vote. In light of their unsubstantiated claims, it is time to confront conspiracy theorists with their own questions. We demand proof that their claims were not motivated by greed for financial gain. We also demand proof that they actually believe in their own theories. Neil Armstrong was correct to counter Bart Sibrel by stating that the Bible on which Sibrel had asked him to swear could be forged, making Sibrel look like a fool.[10] Jim Oberg, a NASA expert on orbital maneuvers

---

[9] This assertion is entirely made up. According to various surveys, only about 5% of US citizens do not believe in the Moon landings.
[10] Sibrel produced the film *A Funny Thing Happened on the Way to the Moon*, in which he describes the Moon landings as a hoax. Professionally and very suggestively, he takes up the typical arguments of the doubters, but works in half-truths or alleged facts that are simply wrong. Even Neil Armstrong is called

and science journalist, described the opponents of the lunar landings as "culture vandals," setting Kant's remarks on enlightenment in a modern context. I don't think we should let our minds or our cultural and scientific achievements be clouded by people who do not value or respect science.

Interestingly, it is usually overlooked that the contradiction between modern enlightenment and technological progress on the one hand, with myths and spirituality on the other hand, is merely a question of interpretation. Meanwhile, there are several remarkable documentaries about the lunar missions that go into more detail about the technical, scientific, and cultural aspects of the lunar landings that can be found online. I mention them here in order to motivate you to delve more deeply into other sources of information, and because these documentaries are more compact than all of the other sources of information you can find; also, they can be easily found on the internet. These wonderful works reflect both the technical, creative, and human aspects of such an extraordinary undertaking. At the very least, the films give you an impression of what goes on behind the scenes in modern space missions, regardless of whether they are manned or unmanned. It does not matter if you were alive to watch the Moon landings as they occurred or not. The challenges, developments, and successes, reproduced in vivid sound and color, provide you with a sense of aesthetics, admiration, and wonder. These documentaries, similar to the cave paintings of our ancestors, will be regarded as our accounts of our own stories for all time.

For All Mankind by Al Reinhard (1989) is a very meditative, almost hypnotic documentary of the *Apollo* flights. A music score written by the well-known composer, Brian Eno, plays a major role in stimulating such a feeling. The film is comprised almost entirely of video recorded by astronauts, and they serve as narrators for the film as well. Their comments, therefore, provide you with first-hand impressions and experiences.

In the Shadow of the Moon by David Sington (2007)—The award-winning film premiered at the Sundance Film Festival not only sheds light on the technical aspects of the American Moon program, but also allows astronauts to provide their own opinions and commentary. The impressions and memories of the men give the film a very personal touch, and their experiences can

---

in as an alleged prime suspect. In a speech to young students, he said that Armstrong even had to hold back his tears by using "cryptic words". Only those who have never seen Armstrong in interviews before could believe the idea that he is "close to tears". This is like the assumption that some people shout just to cover up that their voices would sometimes quiver. Those who don't see conspiracies everywhere, but instead stick to the facts, can hear from Armstrong's speech that he is motivating young people to make new discoveries in the future.

certainly help us be better prepared for future developments in space exploration.

The Moon Machines by Nick Davidson and Christopher Riley (2008)—The six-part television series, produced by Discovery Communications as part of a "Space Week" and broadcast by the Science Channel, looks back at the extraordinary technical problems and developments that were necessary for the flight to the Moon. *Moon Machines* documents the eight-year-long efforts of the approximately 400,000 people involved in the design and manufacture of the new vehicles and equipment. The subjects of the episodes of *Moon Machines* are: 1) Saturn V—2) Command Module—3) Navigation—4) Lunar Module—5) Suits—6) Lunar Rover.

All three films deal with, and answer, a whole series of critical remarks made by the Moon landing doubters and refute them directly. The films also show how conspiracy supporters and their unsubstantiated arguments deny the work of the clever engineers, whose modesty and clarity come without publicity—which very explicitly displays the differences between the mindsets of the engineers and the doubters.

In the end, we can see that it is possible to protect yourself against false claims. Thanks to the extreme flood of information from the media, these counter arguments and protections are urgently needed, even if they do take some time and effort. In any case, subjective speculation and unaudited sources will never be sufficient for examining and evaluating historical events and technical developments. Instead, logic and commonsense will help to understand complex issues. The broad and emotionless dispersion of information sources protects us from charlatanry and obvious nonsense, even in everyday life. This kind of approach is both useful and meaningful in all aspects of life. The astronauts that flew to the Moon were well aware of that, so I'll leave the last word on the subject to them:

- "If two Americans share a major secret, at least one of them will go directly to the press, and there is no possible way you seriously believe that thousands of Americans could keep their mouths shut." (Michael Collins—*Apollo 11*)
- "We've been to the Moon nine times. If this was fake, why would we have faked it nine times?" (Charles Duke—*Apollo 16*)
- "The truth needs no justification. My footprints are on the Moon and nothing and nobody can take that away from me." (Eugene Cernan—*Apollo 17*)

16

Technology, Money, and the Return to the Moon

The Moon remains an object of great attraction to mankind. It has a noticeable effect on our lives (sleep, tides), and many people feel an emotional connection to it. Who doesn't like to gaze up at the Moon at night? After all the discussion about the authenticity of the Moon landings, I am constantly asked whether or not I think humans will ever visit our satellite again. With the extraordinary success of the lunar missions, many people ask why the Americans stopped making trips to the Moon and why the Soviets gave up going there altogether. After all, it is the first stop on the way to the planets, the ever-present goals of all would-be space explorers. And since I myself am helping to shape the German space program, I get asked whether or not we will head back to the Moon in the foreseeable future. In light of the rapid pace of technology development, this question is not easy to answer. Who would have predicted the impact of the internet 30 years ago? Nevertheless, there are indications that allow us to speculate on the future of spaceflight. An unbiased evaluation of the necessary technologies, as well as realistic considerations of the expected efforts and costs, can be an effective way to inform this discussion.

First of all, we have to acknowledge that the extraordinary conditions of the 1950s and 1960s were necessary for spaceflight to flourish. But in the present day, many of these conditions, as mentioned in Chapter 2, are no longer valid. The post-war period and the Cold War are over. The world now essentially revolves around economic conditions, while the opposition of two systems of

© Springer Nature Switzerland AG 2019
T. Eversberg, *The Moon Hoax?*, Science and Fiction,
https://doi.org/10.1007/978-3-030-05460-1_16

economics has given way to sober and cost-oriented planning.[1] Only the mechanisms of politics and public displays of power have hardly changed from before. Whether it is the Moon or Mars, almost every American president has announced grand new plans for space exploration. Unfortunately, political will alone is insufficient for getting new space projects off the ground. Financial and technological conditions must also be favorable.

Back in the age of the stagecoach, respected doctors believed that if a person traveled at speeds over 50 kilometers per hour (31 miles per hour), their lungs would be inflated by the airstream and burst as soon as they opened their mouth. This assumption was proven wrong when the first steam locomotives were developed at the beginning of the 19th century, capable of reaching speeds up to an incredible 60 kilometers per hour. The train conductors wearing their stovepipe hats were considered death-defying heroes, which was not such an outlandish comparison considering steam boilers were known to explode every now and then. You might laugh at these fears today, but you have to keep in mind that, at the time, it was completely unknown as to how the human body would react to such speeds. Every technological step forward is a step into the unknown. We do not have a crystal ball to look into the future, only our rational observation of what happens after new developments. With this in mind, I would never claim to know the limits of future technological advancements. Who wants to make a fool of themselves? Our world has been repeatedly revolutionized through technology and is currently experiencing a massive boom in the understanding of new and exciting areas of science. The miniaturization of semiconductors, nanotechnology, materials research, and genetic engineering, just to name a few areas, are all being heavily invested in by companies trying to earn more money and increase prosperity. The changes in the world caused by information technology alone is probably only comparable to the revolution caused by the printing press over 500 years ago.

The field of aerospace provides a spectacular example. Imagine this: Two printers with the last name Wright, build a rickety flying machine out of wood and fabric, fly a few meters, and then 70 years later people land on the Moon (Fig. 16.1). There are many, decisive steps in history that are contained within the bounds of this sentence. The development of flying metal apparatuses, hissing jet engines, air navigation, exotic materials, fluid mechanics,

---

[1] On 21 October 2011, a Russian Soyuz rocket was launched from the equatorial European launch site in Kourou, French Guiana for the first time in order to achieve a higher carrying capacity and thus reduce the costs of transporting two navigation satellites. In times of the Moon landing, allowing a launch by the "enemy" on your own soil was completely unthinkable.

Fig. 16.1   The Wright brothers' first powered flight on December 17th, 1903 in Kitty Hawk, North Carolina, USA. Photo: www.wright-house.com/wright-brothers

and so on. In aviation, every development was driven by either military or economic interests. Many billion-dollar industries could not exist without air transportation. In manned spaceflight, however, market capitalization by private industry has been limited to bizarre and expensive tourism available only to millionaires, such as flights to the International Space Station or parabolic flights on an aircraft that provide short periods of weightlessness. It is difficult to see any direct benefit to the economy. The situation is different for the unmanned aerospace sector, especially in the areas of telecommunications and navigation. Companies design and develop communications satellites to sell data transfer capabilities. And the market created by navigation systems such as the American Global Positioning System (GPS) extends not only to the navigation systems in cars, but also to the control of agricultural harvesting machines and the intelligent planning of freight traffic. In this case, government initiatives have turned out to be great investments.

Even though past investments have yielded extraordinary gains, the high up-front costs and the less clear return-on-investment still turns out to be a major issue for funding new space projects. Adjusted for inflation, the costs of the Moon landing program in the present day would be around 120 billion dollars. In light of the recent financial crises and the efforts to overcome them, expenditures of this magnitude are actually not as absurd as they may have once seemed. The problem of raising such a large sum of money for space projects, however, lies elsewhere. You have to consider the following: A new mission to the Moon would not simply be about repeating the Apollo landings (why would anyone do that?), but at the very least should include installing a permanently manned station on the Moon. Of course, this would require the development of new technologies. But if you consider comparably complex projects for which new technologies had not yet been invented, the actual costs have never ended up being the same as the forecasted costs. Think of the development costs for the Space Shuttle and, in particular, the International Space Station (ISS). It is very likely that the previously mentioned price estimate for new lunar missions are too optimistic. If you add in the corresponding cost forecasting factors for a mission returning to the Moon, then the cost estimate would certainly be at a completely different order of magnitude than 100 billion dollars. Now, one might object that the costs could be spread over several nations and over many years with international partnerships, but this would not be a deciding factor. The much more interesting question is whether, considering the notorious fiscal deficit and resulting economic crises, space faring nations are at all willing and able to raise funds for an undertaking whose necessity is doubted by the public and even rejected in some scientific circles. The American scientific community has unanimously warned against committing to manned spaceflight at the expense of both scientific and cheaper research missions. And the German Physical Society rejected the pursuit of manned space travel in every aspect as early as 1990 in an official memorandum. Focusing on unmanned missions does not present detrimental effects or pose the same threats that often plague manned missions (i.e. unforeseen costs). Germany, for example, is a world leader in Earth observation not because of its higher technological and scientific competence, but because of its other fiscal priorities.

In choosing what the priorities of the state should be, it is advisable to carefully check whether the people who defend the implementation of expensive projects benefit directly from their financing. Far better, a group of experts from a variety of fields should be consulted for decision making (e.g. the entire scientific community), rather than allowing individual scientists or companies to manage taxpayer-funded missions. Human spaceflight is extremely expensive, and the scientific yield from it has proven to be modest. The conclusion that many people draw from this is that the state simply has to provide more money to make it possible. Clearly, this already implies a decision in favor of manned space exploration, and so anyone who advocates for that alone already presents themselves as ineligible for making an unbiased decision. All states in the European Space Agency (ESA), as well as the United States, constantly struggle with increased national debt, which means that it will remain challenging to finance new space projects in the future.

Looking back once again: When the Apollo program came to an end, it was simply impossible to bring an entire industry comprised of around 400,000 workers to an abrupt halt. The economic consequences would have been devastating. Rather, attempts were made to slowly reduce the number of people in the space industry workforce. More often than not, it was decided to dismiss people who had already been in the workforce for a long time. While this may sound fair and logical to the younger generation, it led to the most experienced engineers being thrown out and the number of failures drastically increasing.

After the extremely expensive flights to the Moon, NASA knew that mission costs had to be greatly reduced (America had to pay for the Vietnam War at the same time), so the old space plane technology from the late 1950s *X15* project was revived. As originally planned, a reusable cargo glider plane would carry the orbiter to great heights, where it could then launch to orbit from, rather than using a disposable propellant tank and disposable solid rocket boosters on the outside of the orbiter. But the concept of a completely reusable spacecraft glider combination became too expensive, and NASA had to switch to the more economical combination that did not use a cargo-carrying glider. The result for the Americans was the Space Shuttle, which was readily marketed as a brilliant technological wonder. This claim got caught up in reality when two Orbiters, *Challenger* and *Columbia*, were lost in two separate disasters precisely because of the trimmed-down design without a completely reusable glider. The Europeans also initially thought of glider technology as positively as the Americans did, designing the *Hermes* space glider and the heavy lift rocket *Ariane V* to support it (Fig. 16.2). Whether or not the Soviets were as convinced of the glider technology is not entirely clear, due to the lack of information exchange during the Cold War. Either way, they developed the *Buran* space glider. Unfortunately, the design goals and predictions regarding the cost and reliability of space gliders were never met. If the Space Shuttle had been as brilliantly designed as it appeared when it was announced,[2] then

---

[2] The Space Shuttle requires solid propellant rockets for the launch. They are considered to be a high-risk system because once they are ignited they cannot be turned off. During the planning phase of the Shuttle, the military warned NASA that, in their experience, they should expect two catastrophic failures per every one hundred launches. This number ended up being quite accurate.

Fig. 16.2 Artistic representation of the space glider *Hermes*. Photo: DLR

we would not have to worry about new crew transportation concepts. However, only the functionality and efficiency of new technologies can be used to measure their success. In the end, the Space Shuttles weren't as incredible as intended, but instead they were quite complicated. Contrary to all forecasts, they were also tremendously expensive: the total price for all development and mission costs up to decommissioning totaled around 175 billion dollars, making the cost per mission around 1.3 billion dollars. Similar experiences with complex (and therefore expensive) systems had by the Soviets with *Buran* and the Europeans with *Hermes* leads one to expect that nothing will change in this respect. At any rate, the *Buran* only flew one unmanned mission in 1988, and the development of *Hermes* was stopped in 1993. Because glider technology has been unable to keep any of the promises made about it (low cost, weekly missions, safety), normal missile technology is back as the primary means of launching manned spacecraft, and private companies are entering the business of launching spacecraft. This makes the main argument for the Shuttle (in 1972 they dreamed of each launch costing 20 million dollars, but the final price per launch ended up being 50 times as large as that!) obsolete and manned space flight remains extremely expensive.[3] Over the years, all glider concepts have been abandoned.

If you express your skepticism about manned space travel, you are repeatedly reminded that without visionary ideas, we never would have gotten to where we are today. That may be the case, but unfortunately such a thesis cannot be examined and therefore remains speculation. What is certain, however, is that we have been able to develop some extraordinary technologies, even though many other technologies failed. But it doesn't help to only look at the technological advancements of the last hundred years. The prophets of a straightforward progression of technology development reluctantly mention that many expectations never become reality. The prosperous future that we were promised with flying cars for everyone and entire cities orbiting the Earth never came to be. Instead, we grapple with home-made problems that we couldn't even anticipate 50 years ago, whose solutions are probably the real challenge facing mankind. In addition to dealing with population growth, this surprisingly includes the issue of "terraforming," i.e. the transformation of an entire planetary environment, as commonly proposed by advocates of colonizing Mars. However, the terraforming that I am referring to is not one that we need to enact on another planet, but rather one that is taking place on our planet, and it is not even fundamentally clear what the effects of it could be besides the melting of the polar ice caps.

All plans about manned missions in deep space, from the Moon to the planets, are still speculative. Anyone who claims otherwise is not familiar with

---

[3] In April 1971 the companies North American Rockwell and General Dynamics presented a calculation for their Shuttle design with a reusable glider plus orbiter (!), which was based on 450 flights until 1988. The total price for development plus operation should not exceed ten billion US dollars.

the present technical, economic, and political landscapes. A few years ago, various lobbying groups in the US spread the idea of a possible lunar orbit in 2017, a landing in 2018, and building a lunar station in 2019. Keeping in mind the typical timelines of projects in the space industry and the experience gained from projects that have already been abandoned, the prospects of the proposed mission concepts were doubtful from the start. The lunar program launched by the George W. Bush administration was cancelled by the Obama administration in favor of more missions in near Earth orbit. Experienced personnel in the major space agencies and companies do not expect to see a Moon landing before 2050, if ever at all.

There are a variety of reasons that are cited to justify flying back to the Moon, and which one you hear depends on which group is proposing it. Astronomers want to build a radio telescope, geologists want to investigate the surface and structure of the Moon, and the space exploration industry sees the Moon as a necessary refueling or staging station for more far-reaching missions into the solar system. A long-running proposal is the mining of helium-3, an isotope which has been deposited in the lunar rock from the Sun. People hope that this can be used as a fuel source for fusion reactors that will solve all of our energy problems once and for all. How lunar helium-3 could technically be mined at all, and what the industrial mining of it would cost, is usually not discussed. The former is unknown, and the latter turns out to amount to fiscal insanity. Instead, such proponents refer to a forecasted increase in energy demand and ignore the fact that fusion reactors running as power plants are not even visible on the technological horizon.[4] Anyone who speculates like this forgets a critical component of the market economy. Products that do not pay off are not developed. The fact that nuclear power plants, bullet trains, and space travel exist at all is due to the taxpayer bearing the risks and additional costs. I think he at least has a right to hear facts, rather than fiction. But if you propose wonderful promises that cannot be kept, then it should not surprise you when the public is skeptical. Let's do the math: The cost of transporting one kilogram (2.2 pounds) of payload into Earth orbit is roughly around 50,000 dollars (in manned space travel). If we generously

---

[4]To justify the necessity of new flights to the Moon with new fuel sources, such as helium-3, is speculative and unsuitable for analysis. There are extensive investigations by competent scientists who prove the opposite (e.g. the report *Factor Four* by Ernst Ulrich von Weizsäcker, Amory Lovins and Hunter Lovins to the Club of Rome) and the claims of a significant increase in energy demand (we are talking about long periods here) are certainly dubious. But I am not getting into the debate of energy savings here. A nice joke about fusion technology is the so-called "fusion constant". Its value: about 40 years. According to the engineers, this is how long it will take to finally develop a functioning fusion reactor. This figure has been constant since the 1950s.

assume that this could be the price of transporting one kilogram from the Moon to the Earth (it is clearly more expensive), then whoever is brave enough to transport ore back from the Moon will realize that the transported ore is orders of magnitude more expensive than gold. To mine profitably on the Moon, the ground there would have to be covered ankle-deep with diamonds.

We can see that both complex technological developments and the decision-making processes associated with them require many years to come to fruition as societies imagine a future world. And this decision making is not only influenced by technical aspects, but also by economic, social, and ecological questions that have to be answered by society itself, requiring extensive public discussion.[5] And, in light of the challenges that face humankind in our present day and age, those questions may even involve our own existence or way of life. When you raise the subject of human space exploration, you would do well to shed light on all the facts that could potentially affect it, and frequently ask people that work in this area about the likelihood of implementation. If you don't, you will tend to lie to yourself about the possibilities. The scientific and economic benefits of human space travel remain doubtful at best. The cost of producing scientific results is extraordinary, and space-based large-scale industry will remain an illusion for the near future. The recurring reference that people make to an "age-old dream of mankind" to explore is more honest, but it remains to be seen as to whether this dream is still feasible in the face of today's financial and social problems. The human race, from *Homo erectus* to *Homo sapiens*, had to adapt to new environments in order to survive through hundreds of thousands of years, and had to repeatedly embark on voyages of discovery either along the coasts or inland. This was the only way that the whole Earth could be populated over the course of thousands of generations. From a developmental-historical perspective, man is an explorer. From a sociological point of view, it would therefore be quite interesting to see how we can compensate for our primordial behavior in a completely discovered, developed, and networked world, if the obvious next step into outer space is prohibitive due to physical and financial reasons. You might as well object here that the next steps of mankind are not predictable. And I agree with you. But the "explorer problem" will be become relevant one way or another. Thinking of the unimaginable and unbridgeable distances to the

---

[5] Obviously, the industrial nations are living well above their means and are now forming financial "safety nets" for entire national economies. The public debt of the USA, for example, clearly exceeds the costs of a Mars mission and one should no longer speak of "astronomical" but of "economic" sums.

stars, man may reach his ultimate limits with the "development of the solar system." A flight to the nearest star, *Proxima Centauri*, would take about 50,000 years with the engine technology available today.

As with the "Moon Landing Conspiracy," the so-called multipliers, i.e. public opinion represented by the media, come into play once again. Here, too, their responsibility as the "fourth estate" is to collect and critically analyze different opinions so that readers, listeners, or viewers can form their own opinions. However, this responsibility is often neglected in favor of market economy constraints—newspapers and private TV channels are profit-driven companies, and spectacles sell better than regular reporting. So, it comes to pass that a simulated mission to Mars, during which some test subjects spend 500 days locked in a closed container, receives far more attention than serious and successful missions of probes to the planets.

Just so that I am not misunderstood: Compared to other challenges of our time, human space exploration is not vitally important. Already, the obscenity of crazy expenditures for weapons or even financial crises in the face of poverty throughout the world can hardly be justified. But in my opinion, it is not about evaluating individual problems of mankind against one another. Rather, in all cases, a culture of discussion should be promoted that considers all challenges and undertakings on their own merits. It doesn't help anyone to just stick to their own field of specialization and ignore or neglect problems in other areas. I work in the management of space projects myself, and I am well aware of the financial and technological constraints, as well as the associated risks of manned spaceflight. I am always skeptical of proposals for manned missions that claim they can accomplish much more than is realistic. We must remind and convince taxpayers of the need for unmanned space missions as a means of technology development to continue making our lives easier and more informed. Today, these technology areas include telecommunications, navigation, and Earth observation from a scientific and cultural point of view (research is a part of culture), along with robotic missions to planets and telescopes for astronomy from Earth orbit. Since this needs to be done in a comprehensible and sustainable way, we cannot afford to fantasize about the deep space exploration of mankind, and we would only hurt ourselves in the long run to build up everyone's hopes and expectations about it. Instead, as with the Moon landing conspiracy theories, critical analysis and open discussions are necessary in order to provide everyday people with the means to form their own opinions (Fig. 16.3).

Fig. 16.3 *Fallen Astronauts* the only work of art on the Moon. It was created by the Belgian artist Paul Van Hoeydonck and left there by the *Apollo 15* astronauts. It depicts an astronaut in a space suit. The plaque created by NASA lists all astronauts who have died in active service up to that point. Photo: NASA. Product No.: AS15-88-11894

Appendix A Apollo Drawings

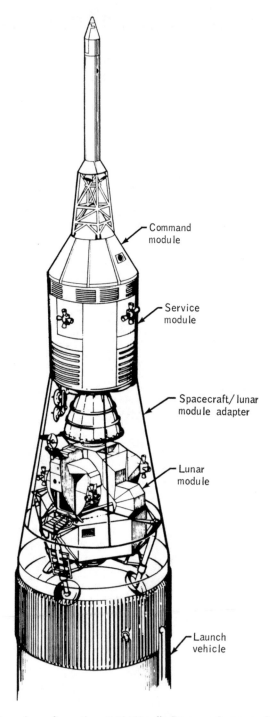

Fig. A.1 Apollo launch configuration. NASA/Apollo Program Summary Report (April 1975)

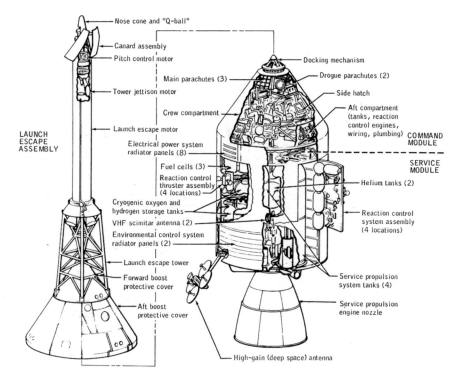

Nose cone and "Q-ball"
Canard assembly
Pitch control motor
Main parachutes (3)
Tower jettison motor
Crew compartment
Launch escape motor
Electrical power system radiator panels (8)
Fuel cells (3)
Reaction control thruster assembly (4 locations)
Cryogenic oxygen and hydrogen storage tanks
VHF scimitar antenna (2)
Environmental control system radiator panels (2)
Launch escape tower
Forward boost protective cover
Aft boost protective cover

Docking mechanism
Drogue parachutes (2)
Side hatch
Aft compartment (tanks, reaction control engines, wiring, plumbing)
Helium tanks (2)
Reaction control system assembly (4 locations)
Service propulsion system tanks (4)
Service propulsion engine nozzle
High-gain (deep space) antenna

LAUNCH ESCAPE ASSEMBLY

COMMAND MODULE

SERVICE MODULE

Fig. A.2 Apollo Command and Service Modules with Launch Escape System. NASA/ Apollo Program Summary Report (April 1975)

LEFT SIDE

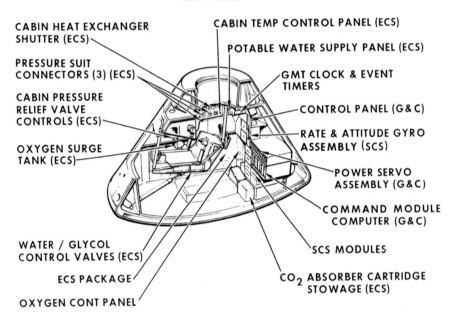

CABIN HEAT EXCHANGER
SHUTTER (ECS)

PRESSURE SUIT
CONNECTORS (3) (ECS)

CABIN PRESSURE
RELIEF VALVE
CONTROLS (ECS)

OXYGEN SURGE
TANK (ECS)

WATER / GLYCOL
CONTROL VALVES (ECS)

ECS PACKAGE

OXYGEN CONT PANEL

CABIN TEMP CONTROL PANEL (ECS)

POTABLE WATER SUPPLY PANEL (ECS)

GMT CLOCK & EVENT
TIMERS

CONTROL PANEL (G & C)

RATE & ATTITUDE GYRO
ASSEMBLY (SCS)

POWER SERVO
ASSEMBLY (G & C)

COMMAND MODULE
COMPUTER (G & C)

SCS MODULES

CO_2 ABSORBER CARTRIDGE
STOWAGE (ECS)

RIGHT SIDE

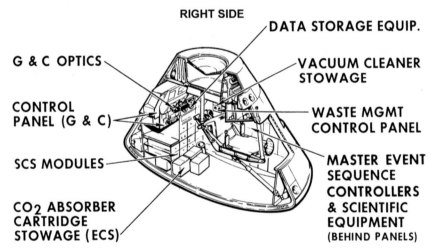

DATA STORAGE EQUIP.

G & C OPTICS

VACUUM CLEANER
STOWAGE

CONTROL
PANEL (G & C)

WASTE MGMT
CONTROL PANEL

SCS MODULES

MASTER EVENT
SEQUENCE
CONTROLLERS
& SCIENTIFIC
EQUIPMENT
(BEHIND PANELS)

CO_2 ABSORBER
CARTRIDGE
STOWAGE (ECS)

Fig. A.3 The inside of the Command Module. NASA/Apollo Training Manual *Apollo Spacecraft & Systems* Familiarization (March 1968)

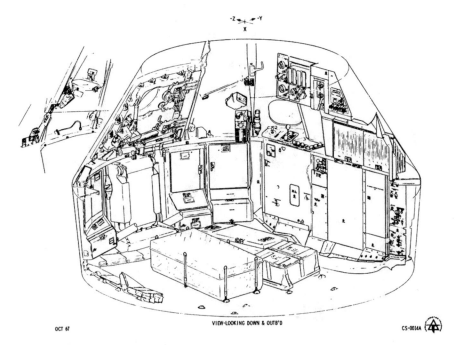

Fig. A.4   The inside of the Command Module. NASA/Apollo Training Manual *Apollo Spacecraft & Systems Familiarization* (March 1968)

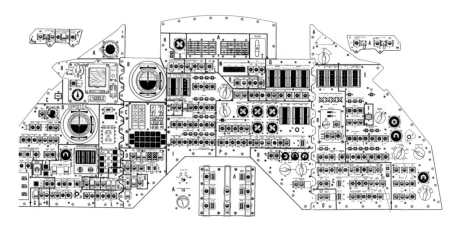

Fig. A.5   Control display unit of the Command Module. NASA/Apollo Operations Handbook Block II Spacecraft (October 1969)

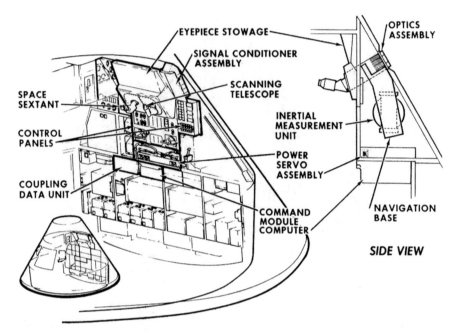

Fig. A.6 Navigation and control system. NASA/Apollo Training Manual *Apollo Spacecraft & Systems Familiarization* (March 1968)

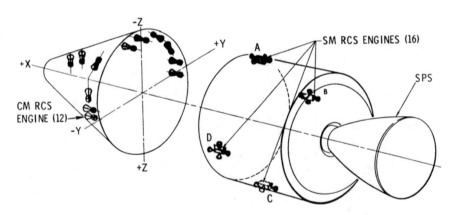

Fig. A.7 Positioning of the drive units on the Command and Service Modules. NASA/ Apollo Training Manual Apollo Spacecraft & Systems Familiarization (March 1968)

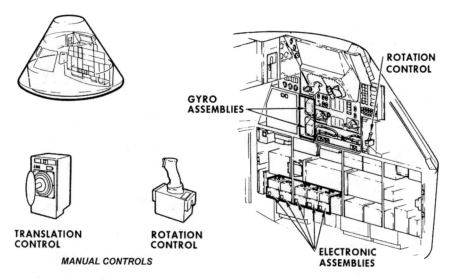

GYRO ASSEMBLIES

ROTATION CONTROL

TRANSLATION CONTROL

ROTATION CONTROL

MANUAL CONTROLS

ELECTRONIC ASSEMBLIES

Fig. A.8 Stabilization and control system. NASA/Apollo Training Manual *Apollo Spacecraft & Systems Familiarization* (March 1968)

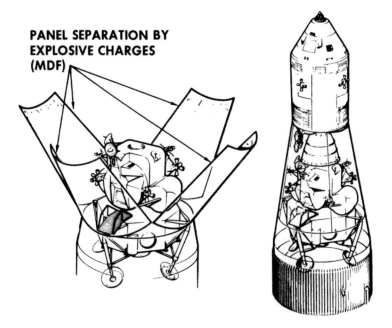

PANEL SEPARATION BY EXPLOSIVE CHARGES (MDF)

Fig. A.9 Adapter for Apollo Spacecraft and Lunar Module. NASA/Apollo Training Manual *Apollo Spacecraft & Systems Familiarization* (March 1968)

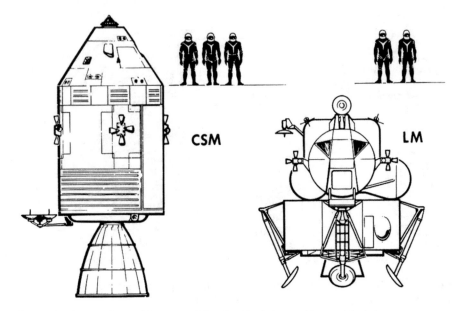

Fig. A.10   Comparison of Command/Service Modules and Lunar Module

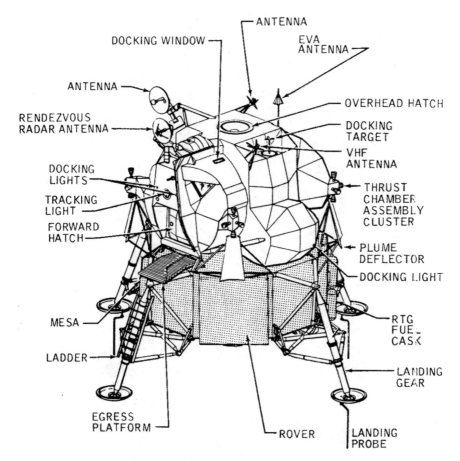

Fig. A.11 Exterior view of the Lunar Module. NASA/Apollo Program Press Information Notebook (1972)

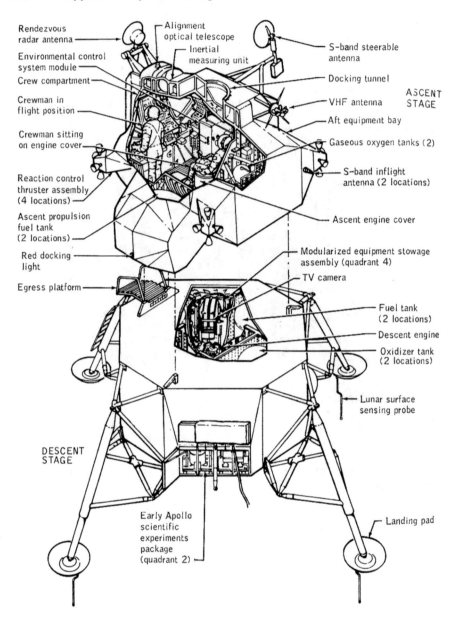

Rendezvous radar antenna

Alignment optical telescope

Inertial measuring unit

S-band steerable antenna

Environmental control system module

Crew compartment

Crewman in flight position

Crewman sitting on engine cover

Reaction control thruster assembly (4 locations)

Ascent propulsion fuel tank (2 locations)

Red docking light

Egress platform

Docking tunnel

VHF antenna

Aft equipment bay

Gaseous oxygen tanks (2)

S-band inflight antenna (2 locations)

Ascent engine cover

Modularized equipment stowage assembly (quadrant 4)

TV camera

Fuel tank (2 locations)

Descent engine

Oxidizer tank (2 locations)

Lunar surface sensing probe

ASCENT STAGE

DESCENT STAGE

Early Apollo scientific experiments package (quadrant 2)

Landing pad

Fig. A.12 Landing configuration of the Lunar Module. NASA/Apollo Program Summary Report (April 1975)

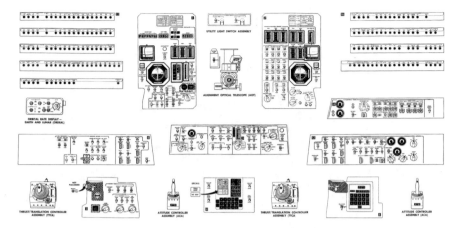

Fig. A.13   Controls of the Lunar Module. NASA/Apollo Spacecraft News Reference

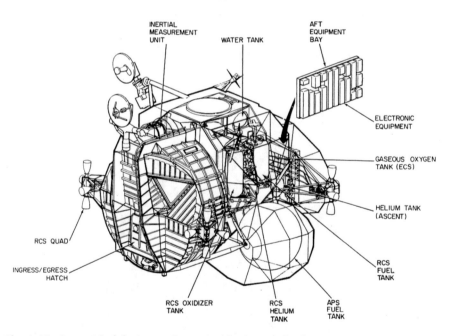

Fig. A.14   Lunar Module Ascent Stage. Inside view to the front. NASA/Apollo Program Press Information Notebook (1972)

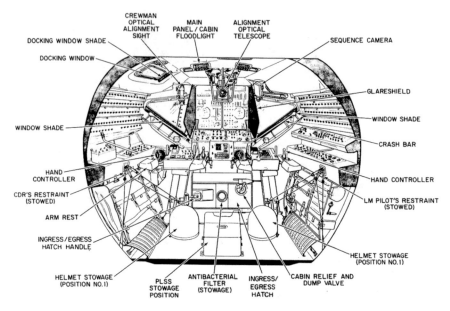

Fig. A.15 Lunar Module Ascent Stage. NASA/Apollo Program Press Information Notebook (1972)

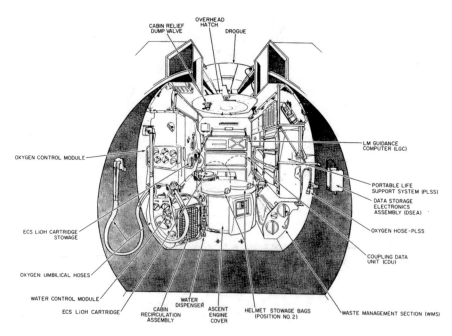

Fig. A.16 Lunar Module Ascent Stage. Inside view to the rear. NASA/Apollo Program Press Information Notebook (1972)

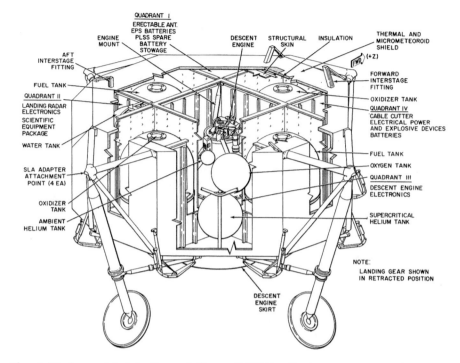

Fig. A.17 Lunar Module Descent Stage. NASA/Apollo Program Press Information Notebook (1972)

Appendix B The Astronauts of the Moon Landing Missions

Fig. B.1 Portrait of the crew of *Apollo 11*. Neil A. Armstrong (Commander), Michael Collins (Command Module Pilot) and Edwin (Buzz) E. Aldrin Jr. (Lunar Module Pilot). On July 20, 1969 the Lunar Module *Eagle* landed in the *Sea of Tranquility*. Photo: NASA. No.: S69-31739

© Springer Nature Switzerland AG 2019
T. Eversberg, *The Moon Hoax?*, Science and Fiction,
https://doi.org/10.1007/978-3-030-05460-1

Fig. B.2 Portrait of the crew of *Apollo 12*. Charles (Pete) Conrad Jr. (Commander), Richard F. Gordon Jr. (Command Module Pilot) and Alan L. Bean (pilot of the Lunar Module). The Lunar Module *Intrepid* landed only a few hundred meters away from the unmanned lunar probe *Surveyor III*, which landed there in 1967. Photo: NASA. Item No.: S69-38852

Fig. B.3 Portrait of the crew of *Apollo 13*, James A. Lovell Jr. (Commander), John L. Swigert Jr. (Command Module Pilot) and Fred W. Haise Jr. (Lunar Module Pilot). A moon landing could not be accomplished, because on the outbound flight to the Moon a supply tank ruptured in the Service Module. The crew was safely brought back to Earth. Photo: NASA. No.: S70-36485

Fig. B.4 Portrait of the crew of *Apollo 14*, Stuart A. Roosa (Command Module Pilot), Alan B. Shepard Jr. (Commander) and Edgar D. Mitchell (Lunar Module Pilot). Photo: NASA. No.: S70-55387

Fig. B.5 Portrait of the crew of *Apollo 15*. David R. Scott (Commander), Alfred M. Worden (Command Module Pilot) and James B. Irwin (Lunar Module Pilot). Photo: NASA. Item No.: S71-37963

Fig. B.6 Portrait of the crew of *Apollo 16*. Thomas K. Mattingly II (Command Module Pilot), John W. Young (Commander) and Charles M. Duke Jr. (Lunar Module Pilot). Photo: NASA. No.: S72-16660

Fig. B.7 Portrait of the crew of *Apollo 17*. Harrison H. Schmitt (Lunar Module Pilot), Eugene A. Cernan (Commander, seated) and Ronald E. Evans (Command Module Pilot). Photo: NASA. No.:S72-50438

Printed in the United States
By Bookmasters